The Economic Reason

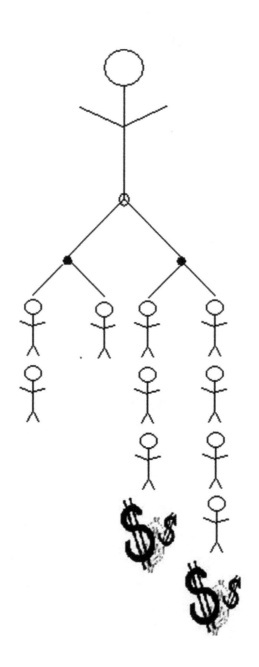

Shane Sanders

The Economic Reason

A Piecemeal Guide to Your Inner
Homo Economicus

 Springer

Shane Sanders
Falk College
Syracuse University
Syracuse, NY, USA

ISBN 978-3-030-56042-3   ISBN 978-3-030-56043-0   (eBook)
https://doi.org/10.1007/978-3-030-56043-0

This Springer imprint is published by the registered company Springer Nature Switzerland AG.
The registered company address is: Gewerbestrasse 11, 6330 Cham, Switzerland

It took me a while, but here it is.

To Dennis Sanders, Debby Sanders, Bhavneet Walia, Simran Sanders, Nanki Sanders, Melissa Sanders, Heather Sanders, Angela Sanders, and Carmen Sanders

Acknowledgments

Thank you to my wife, Bhavneet Walia, with whom I have learned a great deal of economics, and my adviser, Dr. Yang-Ming Chang, without whom I would not have developed many good economic ideas at all. Thank you to my former classmates and current economics collaborators, including (in order of appearance) Joel Potter, Michael Makowsky, Justin Ehrlich, James Boudreau, Christopher Boudreaux, William Horrace, Hyunseok Jung, Merril Silverstein, Shankar Ghimire, Brittany Kmush, Arthur Owora, and others.

I would like to thank Shankar Ghimire for a helpful review of the book. Thank you to Niko Chtouris, Yvonne Schwark-Reiber, Lorraine Klimowich, Chitra Sundarajan, and others at *Springer* for seeing promise in this book and for all their editorial help and expertise. Thank you to all of my helpful department chairs, deans, and senior colleagues over the years, including (in order of appearance) Evan Moore, Ken Linna, the late and great R. Morris Coats, Tej Kaul, Thomas Sadler, Tara Westerhold, Kasing Man, Jana Marikova, Clifton Ealy, Kishore Kapale, Rodney Paul, Mary Graham, John Wolohan, Katherine McDonald, Rick Welsh, and Diane Lyden Murphy. Thank you to my many other friends, including (in order of appearance) Greg Brinkman, Yoshitaka Umeno, Budd Glassberg, John Roberts, Karan Gulati, Nick Klinger, Christopher Vahl, Sailesh Vezzu, Harkamal Walia, Mark Trietsch, Richard Dietz, Brian Jacobson, and Pratap Arasu, for all the high-level conversations. I have learned a great deal from you and other friends over the years.

And finally, thank you to Carmelo Anthony for not shooting a long two-point jumper that one time. That play, isolated from your larger body of work, gave me the feeling that anything can happen in this world.

About the Book

Economics is a far more general paradigm than many people believe it to be. It has been typecast, in a sense, as a theoretical construct with which to study business or large-scale economic conditions. However, economics is the science of human choice and all that definition encompasses. The discipline helps us understand, through a set of assumptions and methodologies, the nature of individual and group decisions. In a series of conversational essays, this book discusses the manner in which economic thought addresses a broad array of everyday issues. In the spirit of the popular economics books of Steven Landsburg, David Friedman, Robert Frank, Daniel Hamermesh, and Gary Becker, to name a few, the book uncovers economic issues and solutions in a decidedly non-technical manner. *Should the federal government mandate use of child safety seats on commercial airlines? Can genetic information substitute for a college degree? Who is served by economic regulation?* The contents of this book touch on many such contemporary topics in an accessible manner.

Unless referring to a specific gender, this book uses the gender pronoun "she" to atone for the imbalanced gender pronoun usage behavior of many a fore-author. Gender references are not meant to narrow the analysis in any way. At its best, economic thought applies to humankind generally, and perhaps beyond...

Contents

About the Author

Shane Sanders is Professor of Sports Economics in the Falk College of Sport and Human Dynamics at Syracuse University. He holds a Ph.D. in economics and has published more than 50 academic articles in the areas of political economy, sports economics, and sport health. Many of these articles have appeared in leading journals of economics, statistics, and sports medicine, including *Journal of Business & Economic Statistics*, *Economics Letters* (4), *Public Choice* (5), *Orthopaedic Journal of Sports Medicine*, *JAMA Network*, *Journal of Sports Economics* (3), *Journal of Quantitative Analysis in Sports*, *Renewable Agriculture and Food Systems*, *Theory and Decision*, *Journal of Economic Education*, *European Journal of Political Economy*, and several others. For his collaborative research efforts, Sanders has received numerous grants, as well as a Finalist Award at the 2019 *Carnegie Mellon Sport Analytics Conference*.

Sanders teaches courses in sports economics and sport analytics at Syracuse University and also serves as a player analytics consultant to professional and NCAA basketball teams. In his teaching, Sanders enjoys the challenge of reducing complex ideas into their simplest elements and then building those ideas back up for students in a relatable manner. Moreover, he believes that economics literacy could be much greater in contemporary societies toward the betterment of humanity and that it is partly the role of economics professors to help realize this potential.

Sanders lives in Fayetteville, New York, with his wife, Bhavneet, and two daughters, Simran and Nanki.

Economic Thinking on the Road (or Thereabouts)

We spend a large portion of our lives moving from point to point. In a series of short, accessible essays, this chapter discusses some of the current economic issues surrounding how we do so.

Policies of the Oscar Wilde Variety

Policies often have unintended consequences. This is a familiar point among economists and students of economics. Since Samuel Peltzman's result that seatbelt mandates lower the effective cost of reckless driving and thus increase the incidence of auto accidents [1], the profession has been on the lookout for perverse effects of even the most well-intentioned government policy. The hunt has been bountiful. Social issues are often akin to bacteria in that their very treatment can lead to unexpected mutations. Of course, unintended consequences can also be fruitful in nature. In the book *Freakonomics* [2], economist Steven Levitt and writer Stephen J. Dubner point out that the legalization of abortion in Romania caused a decline in crime rates 20 years hence. Drawing on earlier academic work by Donohue and Levitt [3], Levitt and Dubner interpret this result as an unintended consequence of the abortion policy. Legalization decreases the price and thus increases the incidence of abortion. As ill-prepared parents are disproportionately likely to select into abortion, legalization essentially limits the relative number of at-risk children being born. Whatever one's personal view on abortion, the economic lesson from this policy analysis is universal. Such policies, whose effect turns out to be more (or less) than meets the eye, can cause targeted problems to persist and untargeted problems to diminish without apparent explanation.

As with laws mandating seatbelt use or those banning abortion, the *Federal Aviation Administration's (FAA's)* consideration of a child safety seat mandate on commercial airlines provides an example of a prospective policy in the mold of Dorian Gray [4] (i.e., apparently beautiful but masking a less appealing core). This very policy issue was evaluated in an academic paper that I co-authored with Dennis Weisman and Dong Li [5]. As the study indicates, it is often difficult to anticipate all effects of a commercial regulatory policy.

History of a Policy

Since the beginning of US commercial flight, children under the age of 2 years have been allowed to ride on the lap of an accompanying adult. This policy, or non-policy really, raised no controversy for many years, as an effective child safety seat did not exist until the 1980s. Following technical advances in that decade, public pressure grew for a child safety seat mandate on US commercial flights. The situation reached a boiling point after a commercial airplane crashed over Iowa in 1990, causing three infant lap ejections. In the aftermath of this incident, *Congress*, the *National Transportation Safety Board* (*NTSB*), and various private lobby groups demanded that the *FAA* institute a child safety seat mandate.

After a long-standing debate that included input from safety engineers, medical personnel, politicians, and economists, the *FAA* announced in 2005 that it would not mandate the use of child safety seats on commercial airlines due to the unintended consequences of such a policy. My study with Dennis Weisman and Dong Li finds that such consequences are present and potentially large in magnitude. Namely, we estimate that a child safety seat mandate would save 0.3 infant lives per year in the air but cause an additional 11.5 deaths per year on the nation's roadways. Thus, the policy is estimated to result in a net *loss* of life. In fact, it is expected that such a policy would cause a net loss of *infant* life. In this setting, the unintended policy effect turns out to be so strong in expectation partly because highway travel and commercial airline travel are fairly strong substitutes. Two goods are said to be substitutes if an increase in the price of one of the goods causes demand for (sales of) the other good to increase. For example, *Coca-Cola* sales rise when the price of *Pepsi* increases, *ceteris paribus*. In the present case, those who choose not to fly due to an increase in price often engage in road travel instead. Further, highways are many times more dangerous than airways. Figure 1.1 shows the discrepancy between US highway fatalities and airway fatalities per 100 million passenger miles.

Even in the terror-stricken year of 2001, Fig. 1.1 shows that air travel was many times safer per passenger mile than highway travel. Thus, a policy that makes commercial airlines safer for infants but also more costly to families has the net effect of putting travelers into harm's way. In informal discussions, Weisman has further considered the effect of heightened security measures in United States airports. Of course, these measures benefit citizens of the United States in the sense that they decrease the likelihood of a terrorist attack. However, they also increase the cost of commercial airline travel and therefore cause many travelers to substitute toward relatively risky highway travel. If implemented in excess, airport security has a similar potential to do more harm than good.

In the same vein, the late economist R. Morris Coats discussed popular policy recommendations during the early spread of swine flu (i.e., when we thought the

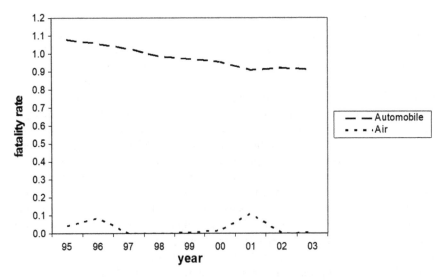

Fig. 1.1 Fatalities per 100 million passenger miles by transportation mode (USA) . Source Data: US Department of Transportation, Bureau of Transportation Statistics, *National Transportation Statistics*, annual [6]

disease to be more deadly). At the time, several legislators encouraged individuals to forego commercial airplanes and other closed-air environments. Coats pointed out that such knee-jerk statements by people of influence raise the perceived (health risk associated) cost of air travel and, as intended, cause fewer people to fly. This may have the effect of saving lives by reducing the spread of disease. However, many of those choosing to forego air travel will once again take to the decidedly less safe roads. This is not to say that it is never a good idea to discourage commercial airline travel. Rather, it is to say that the benefits from doing so should be expected to offset the costs before such a recommendation is ever made. The relative danger of road travel should never be lost on a policy discussion involving transportation issues.

Economic goods are often related. Therefore, a regulatory policy placed directly upon one market almost inevitably affects outcomes in other markets. In general, these indirect effects must be considered if one is to predict the outcome of a given policy.

Traffic Citations: Road Safety Regulation or Municipal Revenue Source?

What is the overriding goal of a local police officer when she patrols traffic? Is it simply to improve road safety? Economists Michael Makowsky and Thomas Stratmann [7] show that the local police officer may have less benign considerations in mind. Makowsky and Stratmann consider all cases in which a vehicle was stopped in a municipality of Massachusetts during the year 2005.[1] Once a vehicle was stopped, they find that out-of-town drivers were more likely to be ticketed for the same offense. The authors attribute this result to the notion that officers are "agents of revenue-maximizing principals." That is to say, local police officers work for elected officials who seek to raise municipal revenues via a tax on non-voters. In addition to pleasing the higher-ups, police officers also seek to make their job easier. Sampled officers in the study were more likely to impose a traffic ticket upon drivers living relatively far from the local courthouse. Makowsky and Stratmann explain that officers do not like to appear in defense of a contested ticket. In addition to the time cost, a court appearance may reveal an officer's procedural error. These results tell us that, for a given offense, the person who lives 80 miles from town is more likely to receive a fine than the person who lives 40 miles from town, and the person who lives 40 miles from town is more likely to receive a fine than the person who lives in town. Makowsky and Stratmann also find that local officers in towns that are (self-) constrained in their ability to raise property taxes are more likely to issue a ticket during a given traffic stop. The same is not true of state police officers operating in these same towns. As state officers are not beholden to town governments, they have no direct incentive to respond to municipal budgetary concerns.

Makowsky and Stratmann uncover many of the factors that influence the choices of local patrol officers. One additional consideration might be the substitutability of a town's roadways. A friend of mine has received only two traffic fines in his life—both in the same small Indiana town. What the town lacks in speed tolerance, it makes up for in directness of route. There is simply no way to get around the town without incurring an additional 15 minutes of driving time. In light of the associated risks, my friend continues to pass through the town in the interest of saving hours of driving time per year. If drivers must pass through a particular town or drive tens of miles out of their way, we would expect them to be more accepting of traffic citations levied by the town's police officers. That is to say, such towns should be able to charge a higher expected "toll," in the form of citations, given the higher inconvenience cost of bypassing them. Conversely, we would expect drivers to be less tolerant of traffic fines in cases that they can easily substitute to roads just outside of the town. It is expected that such towns are less able to raise revenues via traffic citations for a given flow of traffic.

In a related study of Massachusetts property value assessment [8], I teamed with Makowsky to study whether town-appointed assessors estimate property value differently than assessors who are elected by the townspeople. Makowsky and I find that town-appointed assessors raise assessed property values at a higher rate when their town is experiencing fiscal distress, whereas elected assessors do not. In other words, town-appointed assessors act as bureaucrats serving the budgetary interest of the municipal government that appointed them. On the other hand, elected

---

[1]Boston was excluded due to its size and the relatively disconnected nature of its police force.

assessors do not serve town budgetary interests. Incentives matter, and elected officials have fundamentally different incentives than appointed officials who hold the same office. Incentives are often based upon where one's bread is buttered.

Used Car for Sale: Good Condition, 94,000 Miles, Low Variance of RPMs

The modern world brims with measures. This second, I could tell you dozens of quantitative things about myself, not all of which I completely understand. Humans value measures as an efficient means of summarizing potentially useful data. Some measures allow us to achieve consistency (*add 1/2 tablespoon garlic powder*). Others appeal to our need for competitive hierarchy (*My school is better than your school. The Princeton Review ratings say so*). Still other measures allow us to achieve longevity (*My blood pressure is 118/76...My car has accumulated 2,492 miles since its last oil change*). Used car markets function, in the sense that used cars are bought and sold, largely due to various measures that allow the quality of a used car to be assessed, albeit imperfectly. For example, the *Kelly Bluebook* value of a car is predicated upon the number of miles it has been driven, as well as the car's make, model, year, and whether it has endured certain events.

Through the years, the potential used car buyer has come to enjoy more and more information. With companies such as *CarFax* and *Kelly Bluebook*, it is easier than ever to obtain a third-party history or an objective third-party valuation of a used car. Economic theory informs us that society benefits, on net, whenever product quality information becomes public. We are better off for this in the sense that additional mutually beneficial transactions take place as consumers become more symmetrically informed. In particular, Nobel Laureate George Akerlof showed [9] that markets in which sellers know much more about the quality of a product than buyers can cease to operate entirely. Such an informational asymmetry leads buyers to assume that a particular product is of average quality within the market. This assumption, in turn, discourages sellers of higher quality products from participating in the market and drives downward the average quality of products in the market. This process iterates until the average quality of products in the market becomes potentially quite low. Thus, informational asymmetries can cause a market to function below its capacity or, in some cases, to cease functioning altogether. Akerlof uses the example of used cars as an underperforming market in which informational asymmetries can cause sellers to withhold high-quality units of a product. In this market, sellers have more information about car quality than does the typical used car buyer. Akerlof shows that this informational advantage can hurt not only used car buyers but also used car sellers if buyers become hesitant. Indeed, every market participant is hurt if a market fails to generate exchanges.

A measure that is sorely lacking in used car markets is one that directly characterizes the manner and setting in which a car has been driven. Presently, a used car description might mention that a particular vehicle has encountered "mostly highway miles." As the average highway mile involves less acceleration and subsequent deceleration, the potential buyer would prefer such a car, all other things being equal. The problem with such a description, however, is that it is not directly verifiable. There is no odometer in existence that reads town and highway miles separately. Thus, aspects of the used car market continue to resemble a game of dice. Society would receive additional benefit from the used car market if each car tracked the variance of its engine's revolutions per minute (RPMs) over all periods of motion. More exactly, if each car tracked its engine's *average* RPM variance over all trips, weighted by the length of each trip, used car consumers could get a very accurate sense of the manner and setting in which a used car has been driven. A high RPM variance would indicate that the car has encountered miles that are, on average, relatively difficult on the car's engine. Such a car is unlikely to have seen a lot of steady highway driving. A low RPM variance would indicate that the car has encountered miles that are, on average, relatively easy on the car's engine. Such a car is likely to have encountered a high proportion of highway miles.

Lastly, it would benefit the used car consumer if each car tracked the *average* of its engine's RPM level over all periods of motion. Such a measure would help the consumer in assessing the extent of first-order demands placed upon the car's engine. The modern car continuously measures and displays RPMs over all periods of engine activity. However, no mass-produced automobile summarizes the history of its engine's RPM rate. Given the increasing role of computer chip technology in automobiles, the two measures discussed could be easily tracked and used alongside mileage and other factors to provide more predictive and accurate valuations of used cars. This, in turn, would further improve societal benefits originating from our market for used cars.

The New Discrimination on City Buses

There is rampant discrimination on city buses in the United States. Fortunately, it is not the overt and despicable sort that Rosa Parks triumphed over several decades ago. In fact, the present type of discrimination is stealthy—most people don't even realize it's occurring—and, for the most part, perfectly legal. City buses in the United States engage in de facto *price* discrimination. Price discrimination is a practice by which a producer sells the same product at different prices. That is to say, a price-discriminating producer sets the price of a good higher for those consumers who are less sensitive to price increases. Such a practice acts to improve the profit levels of the producer but isn't always feasible. In some markets, it is difficult to distinguish between price-sensitive and price-insensitive consumers, and mistaking a low willingness to pay consumer for her countertype may result in a lost sale. In other markets, there just isn't much difference among the set of consumers.

In a number of cases, producers discriminate based upon the consumer's station in life. Movie theaters often charge patrons lacking a student identification card a higher rate, as non-students are typically willing to pay more for entertainment. Other businesses charge a higher price to those who lack senior citizen status. Consumers can also be differentiated according to their degree of information. In a

Delhi street market, a Caucasian person (such as the author) is likely to be quoted prices that are substantially higher than those quoted to native shoppers. In this case, the Caucasian person's skin color betrays a likely ignorance of the extent to which labor-intensive goods are cheaper in a low-wage country. Moreover, the fact that the person has traveled internationally betrays information about their income level. Commercial airlines are notorious for price discriminating based upon a customer's time of purchase. For example, those who book a flight either 6 months ahead of time or at the last minute are more likely to hold inflexible travel plans. As such, they are likely to be less sensitive to price increases for that trip and therefore tend to be charged a higher fare than those booking a few weeks in advance.

Restaurants such as *Subway* price discriminate by recognizing that some customers want a specific menu item, whereas others simply want food of some sort. The latter group is more sensitive to menu price changes, as they view one food as being essentially the same as, or at least substitutable for, another. If the price of a *Subway* sandwich becomes too high, members of this group will have no problem substituting toward home-cooked food or food from another restaurant. Therefore, they are charged a lower effective price through rotating menu specials (i.e., specials that apply to different menu items at different points in time). These specials don't typically apply to customers who want a specific menu item from *Subway*, as such customers are much less likely to substitute away from that item. However, they *always* apply to the food scavenger type of consumer, who is simply looking for a deal on any food. As a food-scavenging friend of mine used to say, "I eat what needs to be eaten."

There are myriad ways in which firms engage in price discrimination, but how is it that municipal buses engage in this practice? They do so simply by failing to make change for customers. Despite a fare processor that efficiently recognizes and accepts coins and paper money, most city buses do not offer change on overpaid fares. Consider a bus system that charges a $1.25 fare on each ride and does not make change for overpaid fares. This payment mechanism is fairly unaccommodating in that it requires individuals to carry both paper and coins in order to avoid overpaying. Given this circumstance, a $1.25 fare represents an effective price discrimination mechanism in that it separates consumers into two types: Those who prepare for the fare by carrying both dollars and coins and those who do not. Price elastic customers will prepare the correct change ahead of time and pay an exact fare of $1.25 each time. Price inelastic customers are unlikely to prepare the correct change ahead of time and, as a result, will often pay $2 for a bus ride. Bus systems recognize this and charge the latter group a higher effective price by slightly disabling the payment mechanism. Though this method of price discrimination is primitive compared to practices in the commercial airline industry, the objective is the same. A "no change" policy allows bus systems to reap greater profits.

Car MSRP as a Safety Feature

The luxury car business is generally thriving. Many popular car models list for prices above $70,000. There are various reasons that a person would want an expensive car—speed, comfort, style, and safety crash rating being among the top. Some luxury car buyers may be harnessing a more Jedi-like force, however. That is, some people may buy these cars to induce more cautious driving behavior around them. For instance, it seems reasonable to assume that many drivers carry less than $70,000 of comprehensive automobile insurance.

Let us imagine that a person, Bob, is driving along the highway in a *2020 Porsche 911* valued at $95,000. Lisa, a naturally aggressive person who has $50,000 of comprehensive auto insurance, is driving behind him. Lisa understands that she will have to pay Bob about $45,000 out-of-pocket should she manage to total his car. Despite her aggressive nature, we can expect that Lisa will give Bob more space than she would otherwise afford the driver in front of her. In general, we expect the incidence of reckless driving in the area around a $100,000 car to be less than that in the area around a $10,000 car. The driver of the $100,000 car has, in essence, purchased additional road safety. The author, for one, certainly gives more space to a *2020 Bentley Bentayga* than to a *2005 Toyota Camry* when driving along the highway. Aggressive driving and accidents do occur around expensive cars. In 2019, for example, Tracy Morgan's $2 million *Bugatti* was sideswiped within minutes of its purchase [10]. On average, however, we expect such outcomes to be less prevalent in the vicinity of an expensive car.

Upon hearing the above argument, some skeptics point out that drivers of expensive sports cars are notorious for driving *unsafely*. However, this does not mean that sports car drivers don't value a state of the road in which *other drivers* are safer when around them. In fact, aggressive drivers may benefit *more* from the safety of others. Of course, there are myriad ways to improve one's road safety. For example, one might buy a large automobile, drive exclusively during the daytime, buy a car with good crash safety ratings, or drive at a manageable speed. In *The Armchair Economist* [11], Steven Landsburg notes that drivers are more averse to hitting a car that holds a baby. Therefore, some parents (and non-parents!) buy additional road safety by posting *Baby on Board* stickers to their car. Landsburg writes,

> Baby on Board signs. . .are intended to signal other drivers that they should use extraordinary care. I know drivers who find these signs insulting because of the implication that they do not *already* drive as carefully as possible. Economists will be quite unsympathetic to this feeling, because they know that nobody *ever* drives as carefully as possible (do you have new brakes installed before each trip to the grocery store?) and because they know that most drivers' watchfulness does vary markedly with their surroundings. Virtually all drivers would be quite unhappy to injure the occupants of another car; many drivers would be especially unhappy if that other car contained a baby. That group *will* choose to drive more carefully when alerted to a baby's presence and *will* be glad to have that presence called to their attention.

Expensive cars, much like "Baby on Board" stickers, signal other drivers to exercise more caution than usual.

Honking In Lieu of Auto Insurance

On the subject of road safety, it is apparent that drivers in many developing countries tend to honk their horns much more frequently and vigorously than do drivers in developed countries such as the United States. In many developing countries, drivers honk to indicate that they are passing or that they intend to move through an intersection. A driver in the United States, on the other hand, generally reserves the horn for cases in which her fellow driver is moving slowly or making a grievous error.

This cultural difference reflects an underlying difference in incentives. Traffic laws in the United States are highly standardized and enforced, whereas the same is not true in many developing countries. Therefore, a driver in the former setting need not signal her intentions in nearly as many cases. Established road lanes, left (right) of center laws, and frequent intersection stoplights leave less room for negotiation when driving in a typical developed country setting. Further, roads in most developed country settings lack heavy pedestrian and bicycle traffic and therefore feature a lower variance of vehicle speeds. This allows traffic to flow more smoothly, *ceteris paribus*. Lastly, a driver in the United States knows that, by choice or law, other drivers are likely to possess liability insurance. She also knows that civil courts in the United States can deal with accident cases in a relatively succinct manner. Therefore, we expect that the American driver puts forth less effort toward self-protection. A driver in a developing country, on the other hand, has less recourse should her car be damaged due to another driver's error. Thus, she often uses the horn as a defensive driving mechanism. It seems that cultures are often not so different but for their incentives.

When Insurance Is Not Fair

As the *Hertz* representative quoted daily insurance rates for the economy car I was to lease, I converted the charges to monthly and yearly rates in my head.

Forty-four dollars a day is about $1,300 per month or around $16,000 a year.

The rental company's full coverage rate was enormous—both compared to the full coverage rate on the car that I own and to the amount of damage that someone holding a restricted use driver's license would expect to cause on the roadways during a typical day. It struck me that rental insurance has a completely different pricing structure than traditional forms of insurance, in which the policyholder maintains the risked asset over a long period of time.

Actuarially fair insurance covers the (risk-averse) policyholder at a rate exactly equal to her expected loss. If I brought a dozen eggs home from the market each Sunday and dropped the carton on the sidewalk every 12th week, on average, then my expected loss when returning from the grocery store would be one egg. In an actuarially fair insurance plan, I would buy the dozen eggs each week along with a single egg as payment to my insurer. In the event of a dropped carton, then, the insurer would come to my door with a new dozen. Of course, actuarially fair insurance is not observed in the private sector because it is costly to administer insurance plans. For example, insurers must investigate claims to protect against insurance fraud. Further, people are generally averse to large, sudden losses and thus often willing to pay more than their expected loss for coverage in a given period. I might be so opposed to going without eggs for a week that I'm willing to pay three eggs every 2 weeks to cover my clumsiness. In such a case, I would be paying the egg insurer a risk premium (amount above expected loss) of one egg every 2 weeks, or fifty percent of my expected loss.

Let us return to the more plausible topic of car insurance. Given that I do about $1000 of damage on the roadways in the worst of years, the *Hertz* full coverage insurance package entails a risk premium in excess of 1,500%. As I am in no way an atypical driver, such a large premium seems prohibitively high, a fact that led me to ponder whether *Hertz* and other rental car companies might be against renters choosing full coverage insurance. Indeed, *Hertz* faces an unintended consequence that personal car insurers don't face. I care deeply about the state of my personal car's engine a year from now. Such concern leads me to drive somewhat cautiously (i.e., to not accelerate like a certain type of 16-year-old male). On the other hand, I care very little about the state of my rented car's engine 1 year hence because I will have no rights on the car then. Such lack of concern causes me to drive more aggressively when in a rental car.

The only way *Hertz* can influence me to drive *as if* I care about the rented car's engine is to price liability insurance at a reasonable rate and full coverage insurance at a ridiculously high rate. Without the shield of full coverage insurance, my fear of damaging the rented car's body creates a derived demand for engine-friendly driving. *Hertz* will let me have my fun only if I pay a full coverage price that reflects all costs, short term and long term, that I incur upon the car under the auspices of full insurance.

Externalities on the Road

You're driving along in the left lane of a multilane highway at 70 miles/hour. You, and seven cars behind you, are about to pass a semi (Semi 1). However, the driver of Semi 1, wishing to continue moving at a rate of 62 miles/hour, assumes the left lane to pass a stream of semis driving 60 miles/hour. It takes the driver of Semi 1 a full minute to pass her colleagues, during which time all eight cars are forced to slow down. What is the problem with this scenario? Basically, this semi driver gains

1/30th of a mile over the course of the minute by assuming the left lane.[2] However, she imposes a collective cost of about 1.07 miles covered by slowing down eight other drivers.

Taking all drivers as having a roughly equal value for distance covered, then, Semi 1's passing decision is socially inefficient. With social welfare in mind, the driver of Semi 1 would politely wait for the stream of cars to pass. This driver, like most drivers, is concerned primarily with her private or individual welfare (i.e., the distance she is able to cover over a given time period), however. Semi 1's driver has imposed a negative externality upon the eight drivers behind her. According to Auburn University's *Glossary of Political Economy Terms* [12]:

> An externality exists whenever one individual's actions affect the well-being of another individual—whether for the better or for the worse—in ways that need not be paid for according to the existing definition of property rights in the society. An 'external diseconomy,' 'external cost' or 'negative externality' results when part of the cost of producing a good or service is borne by a firm or household other than the producer or purchaser.

In the present case, Semi 1's driver is producing miles traveled subject to a particular deadline, as are the cars behind her. This driver's choice to inhabit the left lane imposes a cost on other drivers that need not be repaid, as occupancy rights to the left lane of a highway inefficiently hinge upon establishment of position (i.e., ability to safely enter the left lane) rather than speed of travel compared to other drivers using the left lane. This circumstance causes an over-allocation of left lane usage. How, then, do we enjoy our road space more efficiently? One technology-intensive but increasingly feasible option is to tax users of the passing lane according to the estimated value of the time they cost other drivers (subject to the speed limit). Such a policy would discourage relatively slow-moving vehicles from taking the left lane hostage. That is, Semi 1 would remain in the right lane unless its driver valued use of the passing lane more than those vehicles it would inhibit. The proceeds of the tax could be used to compensate inhibited drivers for their inconvenience such that the external costs of left lane usage are internalized.

Accidents are another source of negative externality on our roadways. An interstate highway accident might create a traffic jam that causes thousands of people to become late for work. The driver who causes the accident imposes a time cost on fellow highway drivers. However, she is not required to reimburse other drivers for this imposition. Hence, there is an over-allocation of accidents on highways. When a person chooses to drive a tad faster or more recklessly, she does not consider the full social cost that this choice is expected to impose upon society. The solution, again, is to tax the at-fault parties to a highway accident according to the estimated value of the time they cost fellow drivers and to then compensate inhibited drivers with the revenue raised. Such a policy would cause each driver to bear the full social cost of driving faster or more recklessly. Further, the program could utilize the same basic

---

[2]This calculation assumes that Semi 1 has only one opportunity to move into the left lane during the minute of interest.

vehicle identification and account technology that has come to allow for electronic highway tolling. With improvements in real-time technology, it is certainly possible to establish more efficient traffic conventions on our roadways.

The Political Economy of Crossing a Road

There aren't many ways to cross a road. One can achieve a road crossing by moving perpendicularly across the road surface, above the road surface, or, in a somewhat less popular manner, below the road surface. In general, road crossings are a straightforward undertaking. However, there is a troublesome intersection in my former town of residence, where the second-largest city street crosses the largest city street. Cumulatively, I have spent hours waiting to cross this intersection. While this time undoubtedly developed my character in some nuanced manner that I cannot place, the persistent swell of traffic around the intersection suggests that an overpass at the crossing would make a lot of sense. Down the road a mile or so, there is a concrete overpass that allows university-owned cattle to comfortably split their time between pasture and barn. Driving under this structure, I have often wondered how much less value the overpass creates than would a similar structure at the aforementioned intersection. The following exercise will bring us to a rough estimate.

If each townsperson averages 1 min/day waiting at our infamous crossing, then the opportunity cost of an overpass at the location is 5950 h/week. Assuming that an overpass lasts 20 years and each person in town attaches a value of $20 per hour (in current dollars) to her past, present, and future time, such a structure would be worth more than $123 million to the townspeople. By comparison, we can place a rough valuation upon the existing cattle overpass. What function does the existing overpass serve? It allows cattle to pass over a road such that the animals need not be rounded into a trailer and moved between barn and pasture. Let us assume that moving the cattle would cost five people two workdays of each week. When these workers aren't wrangling cattle, we can further assume that they carry out other tasks that generate a present value of $20 per hour to the university's Department of Animal Science. Therefore, the labor-shifting value of the cattle overpass during a 20-year period is a whopping $332,800.

Adding in fuel and vehicle depreciation costs, the value of the overpass might rise to $500,000. Lastly, as these cattle are the subjects of various research projects, we recognize an opportunity cost associated with moving the animals by trailer. That is, cattle researchers, who are distinct from the would-be cattle wranglers, have less access to animals that take time to move. This latter cost manifests itself in the form of foregone or delayed experiments. All told, we might place the present value of a cattle overpass at $1 million. Thus, a people overpass would create many times more value than the cattle overpass is creating. Why, then, is there a cattle overpass but not a people overpass in my former town?

The theory of political pressure groups explains, even predicts, the seemingly misinformed public finance decisions taking place around us. This theory says that a small group with a large per capita stake in a political issue often dominates a large group with a relatively small per capita stake. Such an outcome can occur even when the large group's collective stake is far greater. The basic reason for this outcome, as first put forth by the late economist and Nobel Laureate George Stigler, is that information is typically costly to obtain. Voters must sacrifice time and effort to become informed about an issue (i.e., how it affects her well-being and which officials are likely to support her interests). If the expected gains from such an investigation are relatively small, a priori, a voter is less likely to incur a private cost to undertake political lobbying. The theory does well in explaining why, for example, inexpensive products such as sugar enjoy trade protection in the United States. Though such protections raise the price of sugar for perhaps 90% of Americans, the per capita toll on consumers is not very great. On the other hand, a halt to these trade protections would be very costly to the typical domestic sugarcane or sugar beet farmer. Thus, farmers are more willing to organize and obtain government protection.

In our present case of roadway legislation, the people overpass would benefit 51,000 people each to the tune of about $2400 in present value terms. The cattle overpass benefits a very concentrated and organized political pressure group of perhaps 150 researchers, administrators, and large-scale cattle farmers, each to the tune of about $6667 in present value terms. Hence, a member of the *Animal Science and Friends* political pressure group, prodded by relatively large per capita stakes, will apply more pressure toward obtaining an overpass. Alas, my diluted group may never have sufficient incentive to realize the dream of freely rising above and beyond Tuttle Creek Boulevard along Bluemont Street. Given this circumstance, and the fact that I now live 1500 miles away, I should perhaps find a new political issue.

Mirage of the Parking Lot

Special permit parking spaces help physically disabled drivers and passengers obtain the access to public facilities that most of us take for granted. However, these spaces are generally allocated in an inefficient manner, a circumstance that hurts not only non-disabled people looking for scarce parking space but also disabled people themselves. Jerry Seinfeld refers to disabled parking spaces as the mirage of the otherwise full parking lot. Indeed, such spaces are almost always the last to be filled in a given parking area. It stands to reason that these spaces are allocated such that, even in a worst-case scenario, no disabled person is left out of luck. However, in most cases, we are not under the worst-case scenario, and non-disabled drivers must pass up such spaces in bunches. This is inefficient in the sense that these idle spaces have no value to disabled people at the time but are extremely valuable to those non-disabled drivers stalking the parking lot for an empty space. As such, parking regulation often prevents welfare-enhancing economic activity from occurring.

If non-disabled people were allowed to bid for idle disabled parking spaces, disabled people and non-disabled people could benefit alike. For instance, imagine a program that gives each disabled person a transferable parking voucher. This voucher would allow the person to occupy one disabled parking space at a time over a given area for an allotted number of hours per week, where the person's parking schedule and status could be managed via software application. The voucher would be electronically transferable in the sense that the disabled person could allow non-disabled people to bid for temporary use of this voucher. In this manner, a portion of the voucher transfer price would go directly to the disabled person, a portion to the owner of the parking lot(s), and the non-disabled bidder would receive additional consumer surplus (i.e., benefit from the transaction in excess of here bid price). It is important to note that the electronic auctions supporting this program would represent a spot market that allows bidding only after all disabled people have made their parking schedules clear. Therefore, transferable vouchers could be designed so as not to "crowd out" disabled people from parking spaces at the behest of the non-disabled.

With the advent of "smart meters," which can be used to reserve parking spaces even while en route to a destination, the technology needed to support transferable parking vouchers has become quite feasible. We expect that such a policy would lead to more parking options, better parking facilities, and compensation to participating disabled people for helping to form such a secondary market. Of course, this idea can be extended beyond the disabled parking space. Municipalities, companies, and universities could allow those who purchase a parking permit to sublease the permit in the event that they are absent. Such market activity currently takes place informally but is generally discouraged by host organizations. For instance, many parking permit holders are required to report their license plate number so that it can be checked against the issued permit number. Thus, most parking permits are not freely transferable in the market sense.

Similarities Between Road Networks and Information Networks

At peak hours, Internet users are often downloading more bytes of information than is socially optimal. This over-allocation exists because Internet usage is traditionally priced according to connection speed rather than the number of bytes downloaded. It costs me nothing, on the explicit margin, to access a few megabytes during prime Internet usage hours. However, my usage at such times imposes a cost on fellow network users in that an already congested network is made slower to other users. A flat, connection speed based pricing structure allows clients to use the Internet at peak times even if the external congestion cost they impose exceeds the value they derive from using the Internet at such times. Internet service providers are beginning to discuss this pricing inefficiency and its resulting peak-load strain on the network. In response, some providers have introduced consumption-based subscription plans, whereby Internet users pay for each downloaded byte of information. Alternatively, some consumer-based groups, such as *Consumers Union*, believe providers should limit consumer access to bandwidth-intensive applications (e.g., high-definition video-on-demand).

Neither of these changes, if adopted broadly, would constitute a best solution to the problem of peak-load congestion. The overriding congestion issue faced by

Internet service providers today—namely, a network that can become crowded and relatively slow during peak hours—has been addressed by highway authorities for years. Moreover, it was addressed by cellular network providers in the early days of cell phones. With guidance from regulatory economists, a limited number of municipal roadway systems and several early cellular data providers have realized the efficiencies borne out of a concept called *peak-load pricing*. Such a pricing system charges peak-time users a premium rate, as their use imposes congestion externality costs on other users and also drives demand for additional network capacity. With the right peak-load price, a network of cell phones, roadways, or computers can realize its potential value to both consumers and producers. The Internet is not such a vastly different type of service that providers need to reinvent the pricing wheel. Internet service providers would benefit by following the model perpetuated by several other markets that feature network characteristics.

In a peak-load pricing structure, the consumer is charged a relatively high unit price when demand is at or near its peak and a lower unit price during nonpeak times. In the case of networks, this pricing structure makes certain that the user values each download by at least as much as the external congestion cost it imposes on other users. It does so by charging the user a marginal price such that she incurs the full cost of the download. This cost (price) is high during peak times, when congested networks are further slowed by a marginal download. However, it falls during nonpeak hours, when congestion externalities are almost negligible. Early cellular phone users in the United States know all about peak-load pricing. Throughout most of the 2000s, the marginal minute of off-peak cell phone usage carried no monetary cost. However, the marginal peak minute carried a substantial per-minute fee after a certain threshold of "free" peak minutes had been crossed.

During that era, I remember regularly picking up my phone at 9:00 p.m. (and 15 seconds or so for good measure) to call my mom and dad at the beginning of the off-peak period. Cellular phone plans of the early 2000s featured other pricing features that have since fallen by the wayside. As perhaps a carryover from LAN line long-distance call pricing, many carriers charged the user (with peak network usage) for making calls but not for receiving calls during peak periods. Moreover, most cell users of that era could review their monthly calls and texts through a usage table that was sent with each month's bill. The standard table included the recipient number and duration for each outgoing call during the previous month. Those lists were likely discontinued just after teenage cell phone users taught us all the art of the high-volume text conversation (lol ☺). Today's monthly cell phone usage tables would likely be 100-page tombs.

As cell phone networks improved, peak-load pricing fell by the wayside. However, you don't have to be a first-gen cell phone user to have experienced this form of pricing. Anyone who has driven through downtown London on a weekday morning since 2003 is also acquainted with peak-load pricing. Peak use motorists impose external congestion costs on other drivers by forcing other drivers even closer together and thus increasing the likelihood of bottlenecks. Therefore, London charges these motorists a road use premium to atone for these external congestion costs. As a result, fewer low-valuation drivers are on the roads at these times. Many such drivers will shift to off-peak times, and the level of peak-load road congestion will decrease by design.

Alternative solutions to peak-load Internet use fail to match the performance of peak-load pricing. Rationing each user's bandwidth ceiling at peak-load times, for example, does not necessarily allow for market efficient network utilization. Underneath this solution lies the unqualified presumption that a high-bandwidth use necessarily derives less value, per byte, than one or several low-bandwidth uses. An efficient market for Internet services—one that allows the most value to be derived from its existence—makes no presumption about high-end or low-end uses. Such a market allows any application to be launched provided that the value of the marginal launch to the user exceeds its cost to the network. There is no denying that a *McClaren P1* is an expensive car to build. Society could probably enjoy dozens of *Ford Fusions* with the same resource expenditure that is required to build a single *P1*. If price is any indication of manufacturing cost, then this is certainly the case: a *P1* lists for $1.15 million or the price of 50 *Ford Fusions*. Does this mean that we should build only *Ford Fusions*? The answer revealed by consumer behavior is that we should not. Many thousands of consumers forego the opportunity to buy a fleet of *Fusions* in favor of a single *P1*. In fact, I've never heard of anyone owning a fleet of *Fusions*. *P1* owners do not view one high-end unit as equivalent to many low-end units. With similar logic, we can understand why four-star hotels build some rooms as studios and others as spacious luxury suites.

With regard to peak-load Internet usage, we should not necessarily prefer to minimize video streams and maximize plain text webpage loads. While it is true that plain text webpage loads are much less bandwidth intensive than video streams, it is also the case that plain text webpages are incapable of showing us funny things cats do. Even if cat videos aren't your cup of tea, most people would agree that the plain text web of the early 1990s and before was not as interesting an environment from a content perspective. Innovation, and demand thereof, has driven the Internet from an anemic, command-driven network to our premiere informational medium. What was considered bandwidth-intensive use 20 years ago (e.g., downloading a picture file) is easily handled by current networks. If carriers or regulatory parties had chosen to limit access to picture files on the early Internet, would we have a network today that is able to support high-definition video? When priced correctly, rather than suppressed indiscriminately, the Internet applications that people demand today can inform Internet carriers as they build improved networks for tomorrow.

If rationing isn't the answer, what is wrong with the trend toward consumption-based Internet subscription plans? Certainly, such pricing structures are a step in the right direction. However, they fail to recognize that networks such as the Internet experience at least two distinct market phases—one a peak-time market with strong demand and another an off-peak market with relatively weak demand. Peak demand creates a higher equilibrium market price than does nonpeak demand. A single price for a network good is typically not allocatively efficient because it does not equilibrate both peak and off-peak markets. At least one of the two markets will necessarily feature an under-allocation or over-allocation of quantity consumed. For example, it makes no sense to charge midnight drivers in downtown London a congestion price because their presence does not impose any substantial cost on other drivers. Such pricing would under-allocate the midnight roadways of downtown London. However, congestion prices make perfect allocative sense on those same roadways during weekday mornings. Peak-load pricing realizes that networks, such as those that comprise the Internet, feature two distinct markets separated temporally and each featuring a unique equilibrium market price.

The Underdevelopment of Offline Browser Technology

Information networks have an innate advantage over road networks in their ability to avoid congestion. If I visit the Internet in the early morning, I can download and save information for later use offline. Such an action can help me avoid contributing to Internet congestion during peak use periods. If I go out for a drive in the early morning, this action is not as likely to substitute for the rush hour driving that I will need to do later in the day. Traffic congestion is largely driven by immutable forces such as work schedules and game times. In this sense, information is often more storable than transportation. This storability quality can help Internet users avoid use at peak times to the benefit of all network users.

Many of us encounter scenarios in which a reliable Internet connection is nowhere to be found. It may be while visiting an elderly aunt, during a long airplane flight, or in the middle of a peak use period in which information networks are bogged down. Given such limitations, it is difficult to believe that Internet browser programs like *Internet Explorer*, *Google Chrome*, and *Mozilla Firefox* have been slow to promote offline use through features that streamline the procurement and organization of materials for later use. Offline users are those who employ an Internet browser while unconnected in order to view webpages that were saved when an Internet connection was active. Internet browser programs have the ability to save or download a representation of each webpage that the online user visits. These saved pages can be revisited when the user is offline.

Such an offline page content collection is limited in that it features only visited or selected *pages*. For example, let us suppose that Beth is waiting for a long flight. She would like to peruse the pages of *CNN.com* during the flight but will not have online access to do so. Further, the *CNN.com* website consists of thousands of individual webpages. There is no simple web browser function that would allow Beth to quickly download a representation of all current *CNN.com* pages (or even a defined subset), and she is unlikely to have enough time to select and download individual pages of interest. Therefore, she abandons her plans.

Not only would such a storage function be useful, but it would be feasible from the standpoint of laptop hard drive space provided that bandwidth-intensive pages were converted to a basic HTML format. In economics, we observe that individuals typically prefer smooth levels of consumption over time. For example, a consumer would likely prefer to receive one ripe banana delivered to her door each morning as compared to 365 ripe bananas delivered to her door on the morning of January 1 each year. There is simply no enjoyable way to binge 365 bananas before they begin to turn on you. A web browser that is more offline user friendly would allow for smoother consumption of web material to those who, for whatever reason, face discontinuous online access.

References

1. Peltzman, S. (1975). The effects of automobile safety regulation. *Journal of Political Economy, 83*(4), 677–725.
2. Levitt, S. D., & Dubner, S. J. (2006). *Freakonomics: A rogue economist explores the hidden side of everything*. New York: HarperCollins.
3. Donohue, J. J., III, & Levitt, S. D. (2001). The impact of legalized abortion on crime. *The Quarterly Journal of Economics, 116*(2), 379–420.
4. Wilde, O. (2006). *The picture of Dorian Gray*. OUP Oxford.
5. Sanders, S., Weisman, D. L., & Li, D. (2008). Child safety seats on commercial airliners: A demonstration of cross-price elasticities. *The Journal of Economic Education, 39*(2), 135–144.
6. U.S. Department of Transportation, Bureau of Transportation Statistics. (2005). *National Transportation Statistics*, Annual.
7. Stratmann, T., & Makowsky, M. (2008, May). Political economy at any speed: Determinants of traffic citations. In *American Law & Economics Association Annual Meetings* (p. 124). bepress.
8. Makowsky, M., & Sanders, S. (2013). Political costs and fiscal benefits: The political economy of residential property value assessment under Proposition 212. *Economics Letters, 120*(3), 359–363.
9. Akerlof, G. A. (1978). The market for "lemons": Quality uncertainty and the market mechanism. In *Uncertainty in economics* (pp. 235–251). Academic Press.
10. France, L. R. (2019). Tracy Morgan gets in a crash right after buying a $2 million Bugatti. Retrieved from https://www.cnn.com/2019/06/05/entertainment/tracy-morgan-bugatti-crash-trnd/index.html
11. Landsburg, S. E. (2007). *The Armchair Economist (revised and updated May 2012): Economics & Everyday Life*. Simon and Schuster.
12. Johnson, P. M. (2020). A glossary of political economy terms. Retrieved from http://webhome.auburn.edu/~johnspm/gloss/

Economic Signals

2

> *In economics, a signal is an observable attribute that conveys information about an individual. To use a common example, job hiring committees often interpret a college degree as a signal that a candidate is hard-working and disciplined. In this chapter, we explore educational signaling, as well as a few of the more subtle economic signals in our society.*

Is Genetic Information a Substitute for College?

How many times have you heard a person say, "In my job, I don't really use anything I learned in college."? Given the popularity of this sentiment, why do so many people accept years of debt and sleep deprivation to earn a college degree? For those of us who think enlightenment plays a prominent role, consider the words of the title character in *Good Will Hunting*, "The sad thing is, in about fifty years you might start doin' some thinking on your own and by then you'll realize...you dropped a hundred and fifty grand on an education you coulda' picked up for a dollar fifty in late charges at the public library." In other words, 4 years of steep tuition bills, early morning lectures, and policed essay writing is perhaps not the straightest path to enlightenment.

There are two main economic hypotheses regarding a person's educational choice. Human capital theory says that a person obtains education because it makes her more productive, and thus better paid, in her subsequent career. Signaling theory says that higher education, which often lacks direct workforce relevancy, may not change a person's subsequent work productivity by so much (sleepless nights notwithstanding). Despite this, individuals will choose to undertake higher education at high rates as a means to *signal* their innate ability level. That is, a student might obtain a higher education because an extensive academic record effectively suggests to employers that she is intelligent, resourceful, and not easily deterred from accomplishing a goal. She has completed an arduous task that not everyone is capable of completing. From the perspective of the potential employer, this factor

S. Sanders, *The Economic Reason*, https://doi.org/10.1007/978-3-030-56043-0_2

raises the likelihood that she possesses high levels of grit and innate ability. Both human capital theory and signaling theory hold that education is important toward labor market sorting. However, human capital theory says that education is important because it causes productivity, whereas signaling theory says that education is important because it correlates positively with productivity.

In reality, both human capital theory and signaling theory have a part in explaining educational choice. It has been shown that people attend school both for the purpose of skill gains and as a means to signal their productivity type. The downside of the latter motivation is that signaling through education is a socially costly activity. If ability were freely observable, society could eschew the more superficial, signal-generating forms of education and focus upon types of education that more specifically form human capital or else make people happy. (Let us remember that education can also serve as an avocation. Thank you for this option, art history programs.) However, to effectively reduce the steep costs associated with educational signaling, there must exist a lower-cost method of differentiating job market candidates based on innate ability. Conceivably, such a method exists in the form of genetic sequencing.

In 2007, scientist James Watson became the first human to have his genome sequenced and posted in a public forum. While *Project Jim* carried a price tag of $1,000,000, the price of sequencing subsequent human genomes has fallen rapidly due to vast technological improvements. In the early years of gene sequencing, Jonathan Rothberg, founder and chairman of biotechnology company *454 Life Sciences*, said, "...we are on our way to the $10,000 genome and soon the $1000 genome." Dr. Rothberg was correct to the point that one can purchase a DNA test kit for $49 at the time of this writing. With these advances, we can envision a world in which a job applicant is able to voluntarily submit a productivity prediction report based on relevant parts of her sequenced genome. Potentially, employers would view this information as a substitute for additional schooling (e.g., as observed in an alternative job applicant). The market could best decide what genetic information is relevant in predicting an applicant's level of skill and interest. In turn, this information would be used to predict the applicant's productivity in general and, more narrowly, her profession-specific productivity.

In some ways, genetic information has the potential to outstrip educational information in terms of identifying productive people. Educational attainment does well in suggesting a person's innate ability level but says somewhat less about that person's true interests or potentials, which also have a bearing upon productivity. Most college students, for example, might declare one or two majors during their undergraduate tenure.[1] Often, these choices are based upon noisy recommendations from parents, older siblings, teachers, professors, and friends. How many students are actually choosing what turns out to be an optimal field of study and work for themselves? Like a paleontologist looking for dinosaur bones, most of us probably search haplessly for our ideal major and career path and subsequently fail to discover

---

[1]Like me, other students achieve the trifecta of college major declarations.

it. Genetic information, on the other hand, can identify with a high degree of confidence whether an individual is predisposed to enjoy or merely put up with a particular activity.

Certainly, governments should regulate the flow of genetic information in job markets. It is potentially unsavory for companies to receive anything more than information about a person's productivity potential. For instance, a company interested in retiring workers cheaply might look for job candidates who are more likely to suffer a fatal heart attack in their late fifties. However, government regulators could ensure that companies are allowed to receive a narrow, albeit useful, genomic view of each candidate that volunteers said information. Of course, individuals would preserve the option to continue signaling solely through education, but voluntary moves toward gene-based productivity prediction would overcome much of the present social cost of identifying talent in labor markets. It would also allow us to reimagine education as something less tied to the soul-crushing task of resumé building.

Signaling Skills with Computer Code Rather than Genetic Code

In 2006, *Netflix* challenged the public to improve upon the *Cinematch* algorithm for predicting film user ratings. The company offered one million dollars to any entrant(s) who could improve the expected error magnitude of the algorithm by at least ten percent. The challenge sparked interest and toil from hundreds of statisticians, computer scientists, mathematicians, economists, and other such types. Almost 3 years into the contest, teams realized substantial gains by merging their algorithms—a strategy that eventually led to the toppling of the ten percent error reduction threshold. Anyone who has read James Surowiecki's book, The Wisdom of Crowds, will find it fitting that a combination of strong and complementary algorithms, rather than one great algorithm, managed to obtain the targeted improvement.

Without such an openly competitive structure, it would have been exceedingly difficult to generate several, varied algorithms that could be paired with one another in a complementary manner. An in-house R&D effort toward algorithmic improvement would likely have generated a stand-alone "expert" approach, for example. At the time, this contest was a novel mode of conducting research and development. *Netflix* motivated thousands upon thousands of capital and labor hours onto the project by issuing a challenge and a prize of one million dollars. To obtain a ten percent improvement in their algorithm via conventional research and development would likely have cost tens of millions of dollars. Even then, such an improvement would not have been guaranteed.

Can we then blame *Netflix* for exploiting contest participants? On the contrary, the tournament acted as something of a data science talent search for many of the contestants. That is, it provided a unique environment in which *Netflix* and other firms could gauge the ability levels of contest participants. Indeed, many leading participants obtained lucrative jobs and consultancies as a result of their performance. A leading contestant with the username *Just a guy in a garage* was unemployed before and during the contest. Subsequent to the contest, however, he received offers to work for dating services, futures traders, and health insurance companies. The contest called into question the traditional mode of hiring at many companies that rely upon research and development. Just as the music industry opened itself to national talent searches through network shows such as *The Voice* and *American Idol*, companies in other industries are also beginning to find diamonds in the rough by issuing public challenges.

Certainly, it is not always easy to differentiate individuals based on school performance. Schoolwork seldom demands a novel use of skills in the same way that research and development often does. Indeed, universities may prove to be an increasingly outdated means of skill-signaling in the age of the *Netflix Prize*, *Kaggle.com*, and other such massive skill discovery contests.

When It's Good to Be Just Another Suit

Why are formal suits worn at many workplaces in America? Most people I know believe them to be uncomfortable and difficult to maintain; yet they remain a largely unthreatened symbol of our corporate world. In a blog entry titled, "Why I Don't Wear a Suit and Can't Figure Out Why Anybody Does," self-made billionaire Mark Cuban harkens back to his early days of suit-wearing discomfort:

> When I started MicroSolutions I was 24 years old...I didn't have a closet or a bed, but I had 2 suits. I bought both of those polyester wonders, one Grey pinstripe, the other blue pinstripe for a total of $99 dollars plus tax...I wore those babies when it was cold. I wore them when it was 100 degrees plus. I ironed them and when I could I got them dry cleaned...I went 7 years without a vacation to make that company work, but I didn't go a work day without a suit...After I sold MicroSolution I decided that I never would wear a suit again.

Cuban goes on to question whether the business suit has any real purpose. While his anecdote is very entertaining and rightly calls into question the utility of suit-wearing, it seems that suits *do* serve at least one important purpose in business. They signal to a firm, or to a client firm, that a person is willing and able to conform to a company policy without questioning its logic, even when the policy imposes a significant personal burden on the individual. The bulk of traditional corporate actions involve an employee carrying out a predetermined procedure without significantly challenging or deviating from company protocol. A lot can go wrong with a suit. Therefore, the employee without attention to detail or respect for company policy will stand out like a sore thumb. In the *PBS* documentary *Triumph of the Nerds*, for example, an *IBM* employee recounts his first day at the company. During that day, the man was identified and rebuked for not having worn dress sock garters under his suit pants. He hadn't even made it into the building.

Suits do allow for a degree of individuality but within a somewhat narrow range. Business suits range in color along a thin, but not dimensionless, range of the overall color spectrum. Further, a female employee can alternate between pant suit and skirt suit. The male employee might wear a slightly louder tie or break out the pinstripes, but a suit is a suit at the end of the day. Similarly, companies value employee innovation, but it is generally a controlled form of innovation (i.e., subject to the underlying business process that the firm has established). Thus, companies can imperfectly gauge whether an individual will reliably handle detailed company business merely by looking at how they are dressed vis-à-vis the established parameters of the company's culture.

The hypothesis that an employee's dress signals her work traits is a testable one. If it is true, we should observe a lower incidence of suit-wearing in industries for which innovation is more important than respect for historical protocol. For example, one could stand outside a software company headquarters and a commercial bank headquarters in the same city and determine whether the proportion of suit-wearing employees differs in these two sectors. My hunch is that commercial banking, an age-old industry, enjoys less return from innovation and more return from paying attention to past procedure as compared to the relatively burgeoning software industry. Therefore, I believe we would observe from this field experiment a great deal more suit and tie clad employees in the vicinity of the bank.

Life and Death Signaling

Signaling is not exclusive to the corporate world. Something as simple and silly as a military salute conveys a great deal of information to a military officer. Really, a military salute is a funny custom, like something out of the *Ministry of Silly Hand Gestures*. If a soldier places an open right hand to her forehead at just the right angle, then removes it crisply, then her superiors will be pleased. However arbitrary, saluting performance can quickly separate a good soldier from a poor one. The good soldier remembers to follow orders properly in each situation, whether or not those orders make a lot of sense. In a combat situation, the poor execution of orders can spell disaster. Therefore, a soldier who remembers to salute a superior at the beginning of each encounter signals that she is more likely to be a good soldier. She reflects her training by executing orders without fail. Such a soldier might be considered for platoon leadership or be assigned slightly more demanding duties early on based upon little more than such superficial signals of conscientious execution.

It isn't just the ability to remember to salute a superior that signals a good soldier. It is the ability to remember to salute a superior in the right manner. A sloppy salute indicates that the soldier does not pay close attention to detail. Something is lost in her interpretation of the orders. Such a soldier may be more likely to be careless, and a careless soldier can be dangerous to herself and to those around her. A soldier who salutes sloppily signals that she may be lacking in some important respects. Superiors who treat the salute as a signal will give her lesser responsibilities and perhaps more routine manual labor. If I were in the Army, I would be very haphazard in my saluting behavior. 'Tis usually better to peel potatoes at the bunker than to storm the enemy line.

Culture Signaling

Have you ever heard somebody say that a particular movie was not as good as the book on which it is based? If you're like me, it's usually news that there even was a preceding book, and the comment instantly makes you feel less cultured than your conversational partner. Don't worry. In all likelihood, such a comment is often designed to make you feel this way. People spend much time and effort signaling that they should be admired for their cultural or intellectual depth. While sometimes said in earnest, book v. movie comparisons often appear to be nothing more than boorish attempts to signal one's superior cultural and actual literacy. In order for a culture signal to be effective, the signaler must effectively differentiate herself from the proletariat (e.g., me and people like me). The book comment performs just this function in that books are generally much less accessible than movies. They take more time and effort to finish than do movies, and a book is generally absorbed in a more antisocial setting. Therefore, most people would watch the movie and skip the book.

The intellectually deep person, however, can easily overcome the additional psychic costs of reading the book. Their intellectual and cultural well runs deep and benefits them in that they are able to enjoy more profound experiences than mere movie-goers ever could. Then again, the book might actually be better. If book namedropping doesn't exist as a culture signal, however, I'd like to know why the book is *always* better. At your next cocktail party, I encourage you to make a claim that the movie, any movie, was actually better than the book upon which it was based. You may have to hightail it for the door after the awkwardness sets in, but you will have fought the good fight.

Solvency Signaling

When negotiating the terms of a loan, some banks require the borrower to prove that her down payment has come strictly from liquidated assets. At first glance, this seems like an odd policy. Why does a bank care about the origin of a down payment? The money is just as valuable in any case. However, the ability to provide one's own down payment without taking another loan signals to the bank that the borrower is solvent and therefore desirable as a lower-risk client. In other words, a past ability to accumulate assets predicts a future ability to avoid loan default. Conversely, those who cannot produce a down payment from assets are likely to be at greater risk of eventual loan default. She who saves once is likely to save again another day.

Antiquated Signaling?

Leading up to the *2008 Beijing Olympics*, there was much talk about China's extensive preparations for the event. Many Chinese officials and reporters commented on the country's efforts to display recent economic progress by hosting the *Olympics* in grand style. These efforts involved elaborate building projects, the placement of facades to cover some of the less aesthetically pleasing streetsides of Beijing, and an apparent effort to clear the city of indigents. If prosperity signaling was indeed China's goal, I can't help but think of the approach as antiquated. In the period of history before economic data was available—the *Pre-Macroeconometric Age* if you will—countries often signaled wealth and progress by hosting elaborate international festivals. Greek city-states held *Olympic Games* of the ancient variety to display the might of their athletes and the height of their civilization. A declining city-state might have temporarily belied its weaknesses by hosting elaborate *Games*.

Today, measurement of economic progress is less about signals of prosperity and more about hard numbers such as income per person, unemployment rate, poverty rate, debt-to-GDP ratio, and degree of price stability in a given country. A good modern *Olympic* host will be seen, to a large degree, as just that. It is interesting, however, to observe that government officials around the world often behave with at least some degree of cozenage. Just as a proficient cozener might wear a facade of faux designer suits and seemingly gold timepieces to promote an unrealistic view of herself, a nation might adorn itself in a facade of clean streets, elaborate public constructions, and regional crimelessness (as well as homefulness) to do the same on a grander scale. While hosting a memorable *Olympic Games* may not signal a country's strength to the extent that it used to, such an act may remain important for other reasons. For example, some hosts may be interested in showcasing their country as a viable tourist destination or in forming a reputation as a conscientious international citizen. Indeed, there seems to have been no shortage of *Olympic* host city bids in recent years.

Elaborate Weddings and Credible Marriage Signaling

Citizens of Western countries do not reserve the "mulligan" (or do-over) concept strictly for the golf course. Second weddings have become commonplace in our society. It occurred to me some time ago that a divorced person's second wedding is seldom as extravagant as their first. More than once, I've heard the semi-apologetic line, "It's my second wedding, so we wanted to keep things simple." However, I can't recall anyone saying, "It's only our first wedding, so we wanted to pace ourselves."

Why is the second wedding frequently less costly than the first? Perhaps an expensive first wedding is purchased for more than a memory or a good time. A couple, and the families of the couple, may drop a large amount on the wedding to signal their confidence in the ensuing marriage. That is, if a wedding's backers spend an amount that will take them years to finance, the act indicates a conviction that the relationship is stable. Such a signal may be of value in that it helps the newlyweds gain expedited social acceptance within their community. However, such a costly signal is of value to the signalers only if it carries some credibility. In the case of divorcees, any extravagant wedding signals in the past have proven a false marriage omen. Therefore, guests are less likely to take wedding cost as indicative of future relationship stability when history predicts a different outcome. It would appear, then, that the cost of a second wedding is typically lower because signaling motivations are largely absent at that point, leaving people to *laissez les bon temps roulez* without all of the pomp and circumstance.

Counter-Signaling

There are two types of people in this world—those who would root for the *Los Angeles Lakers* and those who would root for the *Los Angeles Clippers*. It doesn't matter where you live, if you've heard of either of these organizations, or if you've even heard of basketball. The two teams are as class symbolic as East Egg and West Egg in The Great Gatsby. The *Lakers* have trophies and Hollywood fanfare. The *Clippers* have cheap tickets and domestic draft beer nights. Our respective attitudes toward these two teams say a great deal about our culture at large.

In reality, the *Lakers* have historically been a collection of really (in 12.5 point font) good basketball players, and the *Clippers* have been a collection of really (in only 12 point font) good basketball players. At the time of this writing, each team plays in the same building of the same city. In other words, the two teams are quite similar in substance. However, substance means little to status-seeking humans. Somewhere along the line, the *Lakers* became a symbol for status. They won with Jerry West in the 1970s and with Kareem, Magic, and Worthy after him. All this winning begat high-powered fans, and high-powered fans begat more high-powered fans. It is not clear that all of them know so much about basketball, but they are there every game. You can see them right on the sidelines, representing a variety of superfan like no other. To be considered a status symbol in L.A., *Lakers* tickets have long been something of a necessity, and the closer to the sweat the better.

In economics, we call such actions as sitting courtside at *Lakers* games a *status signal* in that such behavior is typically less about seeing and more about being seen. Courtside *Lakers* fans *signal* that they can and do associate with high-powered individuals and organizations. For a signal to be effective, it must set the signaling individual apart from others. Therefore, a credible status signal cannot be accessible to all individuals. Courtside *Lakers* tickets are anything but accessible. Through April 2020, average 2019–2020 ticket price on the secondary market was $253 for *Lakers* home games and $83 for *Clippers* home games [1, 2], where this comparison of averages is likely to gravely underestimate the proportional difference for quality seats. This stark price difference exists despite the fact that the *Clippers* garnered more combined victories for several seasons prior to 2019–2020. It wasn't the basketball game but the status game that has continued to drive demand for (ticket prices of) home *Lakers* games.

Despite all of this focus on the *Lakers*, the *Clippers* do exist. I saw them play on *ESPN3* once. Moreover, the *Clippers* have people who attend their games. One might call such individuals *basketball fans*, but one wouldn't be entirely correct. It seems that some *Clippers* fans blend distinct tastes for basketball and counter-culture. Counter-culturalists get a perverse thrill out of paddling upstream in the river of society. They like to do things so completely opposite of societal norms as to give the rest of society the proverbial middle finger. Counter-culturalists engage in what economists might call, "countersignaling." Whereas attending a *Lakers* game might signal that a person is high-status, attending a *Clippers* game might countersignal either a) that a person wants to see some quality basketball and could care less about such labels or b) that a person actively rejects the status game being played. One might take fan countersignaling as a form of rebellion from the oppression of conventional status contest.

The existence of fan status signaling does not imply that a famous person cannot be a *Clippers* fan. However, it does indicate that a famous person who decides to root for the *Clippers* probably has less value in being a visible part of the established status club. Billy Crystal is a classic example of the countersignaling celebrity. Crystal roots for the *New York Mets* in baseball and the *Clippers* in basketball.[2] In terms of status, the *New York Mets* are the New York City baseball equivalent of the *Clippers*. More specifically, the *New York Mets* are to the *New York Yankees* as the *Los Angeles Clippers* are to the *Los Angeles Lakers*. Both the *Mets* and the *Clippers* are the little metro brothers to a dynastic franchise. Crystal is proof that to engage in superficial status games is not necessarily to have status and to have status is not necessarily to engage in superficial status games. The distinction between the *Lakers* fan and the *Clippers* fan is as much about ideology as it is about true status position in life.

Method of Payment Signaling

At bakeries, cafés, and bars across the country, clients have come to face a strange, new transaction fee in the past couple of decades. This, of course, is the minimum credit card purchase. The question is: *Why does it exist?* Many shops will claim it is driven by the cost of running a credit transaction. Typically, this interchange fee comes to about 2% of the transaction's value plus $0.10. However, such a fee structure explanation doesn't seem to tell the whole story—especially in light of the fact that minimum purchase requirements range in magnitude between $0 and $10 (i.e., from no minimum expenditure to a $10 minimum expenditure).

Perhaps we should look at the customers themselves to understand why these minimum purchase requirements exist. Is there a difference between the person using primarily credit in small transactions and the person using primarily cash? For one, these respective customers track their expenditures differently. The typical cash-based consumer withdraws a certain amount and thereafter can easily keep account of what she spends through a technology called subtraction. However, no such form of recurrent tracking is readily available to the credit-based consumer.

---

[2]Alternatively, it is possible that comedians simply find humor in pain. After all, both Billy Crystal and Jerry Seinfeld remain loyal New York Mets fans.

Then, by virtue of their choice to use credit, it appears that credit-based consumers, on average, care less about tracking small expenditures. For a given small purchase, it would follow that a typical credit-based consumer responds to price increases, by decreasing quantity consumed, less so than does a typical cash-based consumer. Taking credit-based consumers as less price sensitive for a given good, we recognize that any form of price discrimination on the part of firms would act to charge this group a higher effective price. In this sense, credit-based consumers are like the bus passengers who do not carry correct change.

As mentioned in the previous chapter, price discrimination occurs when firms are able to classify buyers. Some buyers will be scared away by higher prices, while others will be less affected. Firms engaging in price discrimination act to charge the latter group a higher price than the former group so as to extract more surplus (or marginal profit) from consumers. Discounts to students and the elderly, who are typically more price sensitive than working adults, constitute a prime example of price discrimination. Minimum credit card purchase requirements induce many credit-based consumers to buy more than they otherwise would. In this sense, it is very likely these firm policies were created not merely to diffuse transaction costs. At least in part, they also appear to represent a price discrimination mechanism used to transfer surplus from the consumer to the producer.

A Questionable Lunch Spot

The following short story taken from the chronicles of my existence supports the conclusion of the previous paragraph. In my former town of residence, there was a restaurant located in the food court of the local mall called *The Bourbon Grill*. *The Grill* was quite innovative in that it was always finding new ways to violate the state's health code. Given the impersonal nature of the workers and the establishment's wanton disregard for modern health standards, it seemed like the type of place that was trying to make a few bucks from unsuspecting mall shoppers. It was decidedly not a "labor of love" type restaurant seeking to bring joy to the stomachs of those around. I once bought a fountain *Pepsi* from *The Grill* after reasoning that it is difficult to contaminate pre-sealed canisters of beverage.

It was a lazy day in *Mall World*, and I distinctly remember being the only customer in sight. The drink was prepared, and the cashier informed me that I owed her $1.06. I took out my debit card and tried to hand it to the cashier. Looking at the card as if it were some form of obscure international currency, the cashier said, "I can't accept credit. There is a five dollar minimum on all card purchases." Cashless in Kansas, I looked down longingly at the fizzing beverage in front of me, told the cashier thanks anyway, and began to walk away. Just then, the restaurant owner appeared at the register, seemingly materializing out of nowhere, and said to me, "It's okay. You can use the card."

For a soft drink addict, this was an exciting turn of events. For an economist, it was an interesting revelation. The owner, who is most certainly not a charity worker in his spare time, went out of his way to accept the card transaction *once I revealed my price sensitivity* (i.e., that I did not have cash and would not buy something frivolous just to meet the minimum) by walking away. In doing so, the restaurant owner revealed that he is better off accepting a card transaction that falls far short of the minimum required purchase amount than engaging in no transaction at all. In that case, the cost of the card transaction alone was not sufficient to inspire the minimum purchase policy. However, there are enough card users who are not like me (i.e., who either carry cash or will make a frivolous purchase to meet the minimum requirement) that it is optimal for the owner to price discriminate, in general, by imposing a minimum purchase policy. One might counter that the owner was simply trying to motivate my long-term business relationship. The point is fair enough, and this may be the case. However, I hadn't seen him at the restaurant many times before that point. It was only later that I even realized he was the restaurant owner.

Since that day, I've found that exceptions to the minimum credit card purchase policy are rare. This is likely true for two reasons. First, the actions of the owner on that occasion did not greatly diminish the credibility of the policy, as there were no other customers around. Also, it is not often the case that a store owner is present for a transaction, and it is difficult for the owner to train cashiers in the nuances of a price discrimination policy ("Respect the credit card minimum purchase policy unless the customer terminates the transaction, and there are no other customers around."). The moral of the story: *When faced with a minimum purchase scenario in which the store owner is present and no other customers are around, just (begin to) walk away.* More generally, this story demonstrates that price discrimination is all around us. It is a sophisticated force that often sees us coming.

References

1. SeatGeek. (2020). Los Angeles Lakers tickets. Retrieved from https://seatgeek.com/los-angeles-lakers-tickets#:~:text=Typically%2C%20Lakers%20tickets%20can%20be,an%20average%20price%20of%20%24261.00 .
2. SeatGeek. (2020). Los Angeles Clippers tickets. Retrieved from https://seatgeek.com/los-angeles-clipperstickets#:~:text=Clippers%20ticket%20prices%20on%20the,an%20average%20price%20of%20%24133.00 .

Labor Issues

3

*Labor issues, and returns to various types of labor, decide
how a large portion of our economic pie is divided. This
chapter discusses how we value and allocate labor within a
modern economy.*

In Support of Child Labor (the Good Kind)

By the age of 18, I had held part-time jobs as a newspaper writer, a grocery store deli
assistant, a driving range pro (or was it a driving range ball picker-upper), a tree
planter, a window washer, and a driveway sealer. I loved some of those jobs and
couldn't wait to get rid of others. All in all, however, I believe that this spree of part-
time job exploration helped me immensely in sorting out my talents and limitations.
In fact, this experience has left me with the belief that teenage part-time employment
is both a sign of overall industriousness in the United States (We let our youth try
their hand in things) and also a reason for our prosperity. In 2018, 72% of US high
school students held a job [1]. In many other countries, part-time, experiential labor
force participation on the part of teenagers is not so common. Some of my friends
from developing regions of the world tell me that it is disgraceful in their society for
a middle or upper class adolescent to hold a part-time job. Apparently, many families
in such regions believe teenage employment to be a signal that parents are struggling
to provide for their family. And so many of these families forbid teenage children
from holding a job.

Mass teenage employment not only keeps the cost of a *McDonald's* hamburger
low in the United States but also allows young people to explore their abilities in
ways that school alone cannot. School didn't teach me that I have a knack for
sportswriting but that 35-year-old sportswriters often drive cars that are falling
apart. School didn't teach me that I have a low tolerance for being in the vicinity
of 120 °F asphalt or that I am much more productive without a designated manager
deliberating my every move rather than joining in on the effort. As an aside, I've

never liked the adage: *Those who can, do. Those who can't, teach.* Perhaps it hits a little too close to home for me. The adage is arbitrarily constructed, as it can easily be flipped to read: *Those who can't teach, do.* However, I've crafted a similar adage that pertains to managers. It goes like this: *Those who can, do. Those who don't want to, manage.* Along these same managerial lines, school did not motivate my view that US workers are severely overmanaged, a view that only grew as I became more skilled in my work. Rather, these lessons were learned through early, part-time employment, and they shaped the career choices I have made as an adult.

However, I'm not the only person who has learned in this way. A slightly more successful American than myself named Bill Gates began debugging software and writing payroll programs for companies in Washington as a young teenager. While this work was, in itself, a sign of Gates' irrepressible precociousness, one wonders if his abilities would have been encouraged to the same degree in a more restrictive labor market. It is tempting to believe that our society's willingness to allow a teenager to experiment with skilled labor helped Bill Gates become a renowned innovator rather than a misguided genius. Perhaps he derived the confidence to eschew further enrollment at *Harvard University* and co-found *Microsoft* from things he began learning about himself as a 14-year-old computer programmer.

There are countless other examples of meaningful teenage employment. Before becoming a master of the printing presses, Benjamin Franklin was an indentured printing apprentice to his brother James. Though he learned a great deal from the apprenticeship, Franklin's experience was less ideal than that of Gates, and he fled his violent brother's presses at the age of seventeen. Thus, there is a vast distinction between child indentureship, which still exists in many parts of the world, and exploratory teenage labor. With proper labor restrictions, teenagers can harness their skills in a safe and productive environment. Part-time teenage employment has allowed millions of Americans to enter the permanent workforce more efficiently in that they have already learned what they can and will do as employees. The result is that more workers hold positions for which they are well suited, an outcome that improves sorting and matching in labor markets as well as general well-being.

Of Labor and Capital

In many futuristic depictions of the world, the labor class has been replaced by machine. Such a notion of the future is justified to some degree. Since the *Industrial Revolution*, capital has become increasingly important in most production processes relative to labor. Instead of 100 professors presenting a microeconomics lecture to 4000 students, one of those professors can now teach all of those students with the help of a satellite or Internet feed and the allocation of 4001 or so computing hours. This does not imply, however, that labor is out of luck. Indeed, there are not a fixed number of production processes in our economy over time. In reality, labor can shift to another, perhaps new, production process if it is crowded out by capital in a given initial production process. Political satirist P.J. O'Rourke has said, "Everybody wants to save the Earth. Nobody wants to help Mom do the dishes." Perhaps more labor would help Mom with the dishes if the Earth were already saved (e.g., by an infusion of capital). The point is that labor can always shift from one endeavor to another according to the needs and skill sets of the times. Moreover, there are simply some things machines cannot do as well as humans. For this latter reason, the labor force in some industries will remain quite stable for generations to come.

A prime example of such an industry can be found in the case of live orchestral performance. With amplifiers and double tracking, one musician could easily represent the violin section of an orchestra. Similarly, an orchestra could feature one flutist, one clarinetist, and a single trombonist. By adding in synthesizers, we could do with even fewer human musicians. Given that music technology has improved so dramatically, why has there not been a proportionate change in orchestral size and structure? There are many forces driving this outcome. Among other things, audiences value tradition, "authentic" textured sound, and the natural challenge of unison among sections. Less obviously, large orchestras are still intact because a single group of professional musicians produces a less centralized and often better sound than does a single musician amplified.

If we simply magnify the sound of one professional musician, for example, she might make a slight error on 0.25% of all notes (1 in 400 notes). This amounts to several audible errors within the frame of a concert. However, assume for simplicity that we have ten musicians in each section. Each musician still errs on 0.25% of notes, but these errors generally coincide with flawless play by the nine other section musicians. In fact, if we assume that errors are independent across musicians, a passable assumption if the conductor is doing a good job, then the likelihood that at least seven musicians play a given note correctly is .9999987. In other words, we expect only one note in 770,530 for which fewer than seven of ten section musicians play correctly. The likelihood that fewer than six section musicians play a note correctly is about one in 205.2 million. Even at the strongest odds Las Vegas is known to offer, one would be smart to bet that every note in a concert is played properly by at least a majority of section musicians. The effect is that musician errors are muffled, for the most part, within the framework of a large orchestra. Despite the presence of individual musician error, each *section* will sound almost seamless from beginning note to end note.

Thus, a full orchestra is something of an insurance policy. With one musician, you will get a well-played note or, not infrequently, a noticeably erred note. With many musicians in each section, some notes might not shine to the same degree, but, in general, the collective product is almost always a well-played note. The result is a fairly stable, albeit somewhat thin market for classical musicians. By employing a full orchestra for our entertainment, we take advantage of a statistical result known as the *law of large numbers*. In the present setting, this statistical law says that the average pitch of a set of notes played simultaneously across a set of trained musicians is expected to tend closer and closer to the "true" pitch, on average, as the number of musicians increases. The larger the number of trained musicians, the more dependable is said convergence.[1] This result rests upon the aforementioned assumption that given errors are not somehow "contagious" across musicians. That is, we assume that the likelihood that one musician errs on a given note does not depend upon whether her colleague errs on the same note, and *vice versa*. One need not visit the symphony to observe this statistical phenomenon. Convergence upon a true note is apparent in church settings, in which congregations of average singers usually sound far better than a single average singer. All of those sharp notes and flat notes will usually balance each other out toward an impressive overall sound.

Of course, it is becoming increasingly true that a musical synthesis simulation program and a musical composition artificial intelligence algorithm could replace all of the musicians along with their composer. It remains to be seen whether music patrons will view such an arrangement as music, however. Part of live musical performance is the mastery of skill over error. Perhaps paradoxically, this mastery may be appreciated only if error is a distinct possibility.

If You're Back Here, then Who's Flying the Plane?

Another "cockroachian" form of labor is that of the commercial airline pilot. Airplanes have flown remotely for years. Therefore, it is clear that we *could* fly commercial airplanes in a more capital-intensive manner. An air traffic controller could program and monitor several flights simultaneously via computer system. Indeed, this is how U.S. military drones provide precise surveillance and payload delivery.

However, we are unlikely to see such a change on commercial airlines any time soon. You see, a pilot on board is a signal from the airline that it is committed to safety. Two pilots are an even stronger signal. By hiring onboard pilots, the airline is contracting skilled employees to effectively post a bond equal to the value of their lives. This bond is typically redeemable only if all passengers survive the plane ride. That is to say, the pilots of a plane are unlikely to survive a plane ride unless all passengers have also survived. This is true because soft commercial plane crashes are somewhat rare. If each pilot values her life at $8 million dollars, then the cost of a crash to those with some control over safety is quite high.

Hence, passengers can be sure that the safety interests of airline personnel are aligned with their own safety interests. In the case of a remote-controlled commercial flight, the airline would no longer have skin in the game. Literally. It is therefore unclear how much effort a remote pilot will put forth in responding to a critical engine problem. Though it would undoubtedly feel like a huge effort for the remote pilot, would it match the effort put forth by a live pilot in the middle of a life and death scenario? In an era when many of us have come to anticipate rapid change, there are countless examples of industries in which labor markets are quite stable. As

---

[1]The central limit theorem further specifies that the "average" such note is expected to vary normally, or in a bell-shaped manner, about the "true" note.

inevitable as it sometimes seems, even change requires a compelling underlying stimulus.

Robot Scientists

Over the past decade or so, scientists have been at work creating sophisticated robot scientists. These robots do more than just arduous tasks within the lab. They have the ability to conduct experiments, analyze results, and vary experiments based on prior results. Robot scientists in the United Kingdom and United States have already developed key results that have eluded human scientists for decades. In early 2009, a UK robot named *Adam* was able to identify an important gene in yeast. Human scientists had pursued this discovery for almost 50 years. In the United States, a robot scientist at *Cornell University* independently replicated the key physical laws of pendulums and springs through experimentation.

Robot scientists are good for science. However, it is unclear whether they will be good or bad for a bench scientist who wishes to continue practicing science. Some see them as powerful laboratory assistants that will allow human scientists to delve more deeply into scientific theory and experimental design. However, robot developers are well on the path toward teaching robot scientists how to conduct experimental design based on an artificial understanding of scientific theory. It is conceivable that science will eventually retain a workforce consisting of only the best human scientists as principal investigators, complemented by computer programmers and highly effective robots.

Economists with Job Principles?

There are essentially five factors that determine the wage of a particular worker: the level of demand for the good or service that the worker produces, the skill requirements of the profession, the training costs necessary to enter the line of work, the desirability of work conditions within the profession (including the presence of job risks and other similar considerations), and the presence or absence of artificial barriers to entering the profession. Among these, the desirability of work conditions has an often fascinating and pivotal influence upon the career choices of individuals and therefore upon the wage levels that obtain in various professions. As a case in point, we can examine the labor market for economics professors.

Economics professors are sometimes employed in business schools and sometimes employed in humanities-based economics departments. Economists John Siegfried and Wendy Stock [2] find that business school economists make 26% more than their humanities-based compatriots. Much of this premium appears to hold even when controlling for the overall quality of the employing university. But what is the basis for this premium? Undesirable work conditions seem to be the main factor driving this premium. Most economists appear to enjoy teaching economics to future economists, researching economics as a social science, and being surrounded by colleagues who can directly appreciate their work. In a 2002 *Chronicle of Higher Education* article, Jennifer Jacobson reports [3]:

> In each of the past three years, *Stanford University's* business school has gone head to head with its economics department and lost. What have they been fighting over? Faculty members with PhDs in economics. David M. Kreps, senior associate dean for academic affairs in the university's business school, says it has lost junior and senior scholars to *Stanford's* economics department, even though his school offers them salaries that are $20,000 higher.

Stanford University's experience provides us with a fairly clean natural experiment, albeit one lacking in sample size. For a given university (i.e., conditional upon the quality and location of the employer), the economics professors in this example must be paid a handsome premium—one greater than $20,000 in 2002 dollars or about $30,000 today—in order to join a business school. The evidence from *Stanford* is consistent with the broad labor market finding of Siegfried and Stock that economists command a 26% salary premium in business schools, *ceteris paribus*. If, for example, the *Department of Economics* at *Stanford* were offering prospective faculty members a salary of $200,000, then the finding of Siegfried and Stock suggests that the typical economics faculty candidate at the school would require 26% more than this, or an additional $52,000, to work in *Stanford's Graduate School of Business*.

According to several economists working in the area of public choice theory, James Buchanan's 1983 move from *Virginia Tech* was almost certainly prompted by a perceived lack of appreciation from his business school colleagues. Buchanan relocated to *George Mason's* humanities-based *Department of Economics*, taking the *Center for Study of Public Choice*, Gordon Tullock, and several other faculty members with him. It appears from these anecdotes, statistics, and outcomes that the average economist must be paid a premium, in money or favor, to compensate for the relative undesirability of a business school setting. Apparently, some economists do have a bit of principle, a fact that leaves business schools to dig deeper into their pocketbooks.

Labor Supply and Quantity of Labor Supplied in the Immigration Debate

Immigration is a contentious topic. Rather than add to the contention, I wish to clarify a poor economic argument that is frequently used in immigration policy debate. Many people support immigration by stating something along the lines of: "Americans aren't willing to do the jobs that immigrants are willing to do." There are many reasons to support immigration, but fear of jobs undone is probably not one of them. Particularly, this fear is potentially erroneous in that it confuses the notion of *quantity of labor supplied* with that of *labor supply*. Whereas quantity of labor supplied is the amount of labor a person or group provides at a given wage, labor supply is a schedule specifying a person or group's quantity of labor provided at

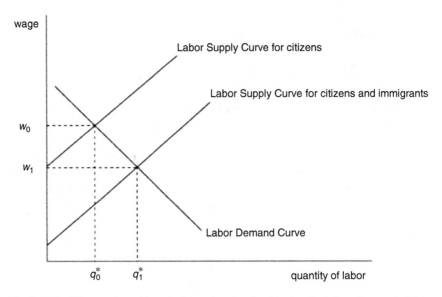

Fig. 3.1 The labor market with and without immigration (Author depiction of a standard labor market setting)

each possible wage level. The commonly held sentiment quoted previously stems from an observation: *that some US jobs are held almost exclusively by immigrant workers.* From this observation, a fallacious conclusion is drawn: *that native workers would not work in such jobs at any wage level.* To say that a group's quantity of labor supplied is zero for a given job and wage is to say nothing about the group's behavior at higher wage levels. Figure 3.1 represents a possible depiction of a labor market with and without immigrant labor.

Figure 3.1 features two supply curves, one showing quantity of labor supplied by native workers at each possible wage and another showing quantity of labor supplied by the combination of native workers and immigrants at each possible wage. In the absence of immigration, native workers would earn a wage of w_0 per hour and work q_0^* hours in the featured labor market. However, if immigrants enter the country and work in the same market, the supply of labor shifts right (quantity of labor supplied increases at any given wage). In equilibrium, workers in this market now earn w_1 dollars per hour and work q_1^* hours. However, by looking along the labor supply curve of citizens, we find that native workers supply zero hours of labor to this market at a wage of w_1. Given their work and nonwork alternatives, native workers opt out of the type of labor depicted in the presence of immigration (and its wage-depressing effect). However, citizens will rejoin this workforce if the wage rises above w_1.

In a labor market welfare sense, immigrant laborers clearly benefit other members of our economy (e.g., consumers and producers). However, it is not the case that jobs would necessarily go undone without them. Rather, in the absence of immigrant laborers, relevant market wages would move upward and eventually induce the participation of citizens. If you don't believe that this response would occur, consider the following series of questions:

> Would you work in a crop field at a wage or $12 per hour? At a wage of $22 per hour? At a wage of $32? At a wage of $42?...At a wage of $222?...At a wage of $1,212?

Unless you are a *Fortune 500* CEO or are severely allergic to plant pollen, there is a good chance that you opted in somewhere in that chain of questions. I, for one, would happily hit the fields for $100 per hour. This thought experiment demonstrates the point that we cannot know what market participants would do in the counterfactual. When confronted with such logic, it is my experience that those who confuse terms like *supply* and *quantity supplied* will usually have no qualms about trivializing the distinction. Such individuals might cast aside the distinction by saying something like, "That's what I meant." However, we cannot set the rules about what particular concepts mean and whether or not they are interchangeable. In economics, each given concept has an objective meaning. With this meaning and no other, it contributes a unique piece to the puzzle of economic thought. However, we must use each such piece as it is designed in order to complete the puzzle.

Perspectives on (Un)employment

In his book, <u>The Armchair Economist</u>, Steven Landsburg points out that unemployment, in and of itself, is not necessarily a bad thing. He reasons that unemployment can be an optimal choice for some people, as it is associated with such economic goods as increased leisure and decreased work effort. Landsburg writes,

> Journalists like to use the unemployment rate to indicate the overall state of the economy. The surrounding discussion usually overlooks the fact that unemployment is something to which people aspire. The leisure to be idle or to pursue one's fancies is generally thought of as a *good* thing...Of course, unemployment can be accompanied by bad things, such as diminution of income, and these are the things that reporters have in mind when they suggest that unemployment is undesirable. But it is worth remembering that the benefits of unemployment help to alleviate the associated costs.

Landsburg is correct that unemployment "can be accompanied by bad things" but that the state of leisure, in and of itself, is often desirable. In this sense, unemployment is perhaps less harmful than commonly reported. There is a related reason that increases in cyclical unemployment (i.e., unemployment that varies with business cycle fluctuations) may be less harmful than advertised. Assume that the unemployment rate increases by one percent due to a cyclical downturn. A typical person might see this rise and assume that each newly unemployed person (a) values job continuity to a typical degree and (b) held a typical set of job expectations when initially taking their previous position. However, one must remember that the type of person most likely to become cyclically unemployed is one in an industry with a cyclically volatile labor market. For example, the level of consumption in an economy is more stable from quarter-to-quarter than is the level of investment activity. Thus, we would expect that a shift manager at a retail store is more immune to cyclical unemployment or underemployment than an average, nonunion brick mason. Workers in volatile industries select into their job with at least some preexisting knowledge of its labor market. In this sense, those people most likely to become cyclically unemployed are often the same people who are most accepting of cyclical unemployment. Such individuals are expected to care less, *ceteris paribus*, about a smooth income stream than workers in other industries.

Furthermore, it is important to note that workers in volatile industries are paid a premium that is based directly upon the inconvenience of their higher potential for layoff. This premium arises because some workers are averse to entering a volatile labor market, thus making the supply of workers competing for such jobs scarcer and driving up the prevailing wage in that market. That is, employment volatility in an industry affects job desirability, labor supply, and compensation within that industry. We would expect workers to select out of a volatile labor market until the wage in that market fully compensates remaining workers for the costs associated with job insecurity.

When considering changes in the unemployment rate, we should remember that unemployment means different things to different people. That said, the unemployment rate is a telling parameter only if we take the time to understand from where new unemployment arises. If symptomatic of a normal business cycle, then rising unemployment has likely been implicitly compensated through prior wages. It is the unexpected, and therefore uncompensated, cases of new unemployment that can claim true economic victims.

Employment Across Country

Fareed Zakaria, former host of the *PBS* television program *Foreign Exchange*, took on a challenging assignment for the show when he attempted to reason his way through the work decisions of Americans as compared to those of Europeans. My takeaway summary of Fareed's explanation goes as follows:

(a) Americans work more than Europeans.
(b) As we work more, our productivity declines.
(c) In many European countries, work weeks are shorter and vacation time longer than in the United States.
(d) In some of these same countries, productivity is higher than in the United States.
(e) Therefore, these countries are better off than the United States.
(f) Thus, US workers and the US economy would be better off if everyone worked less.

My apologies to Fareed if I misunderstood any part of his argument. However, let us continue with my interpretation. Fareed is right on with articles a, b, c, and d. Most importantly, it is generally accepted that worker productivity declines as we spend more of our time working. This is true for a number of reasons. The workaholic real-estate agent, for example, can do many important things early in the workday, such as make phone calls and show houses. However, her options diminish as evening comes. At this point, she might turn to less important tasks such as organizing documents. Also, the longer her work week, the more hours she spends in a state of mental or physical fatigue. These factors combine to make the first work hours she chooses most valuable and the last least valuable.

Despite the early success of his analysis, Fareed takes a few logical leaps in the two concluding thoughts. Productivity is not a measure of well-being. Rather, productivity measures the rate at which goods and services are produced per unit of input within an economy. To say that one economy has a higher productivity rate than another is not to say that the people in the first economy are doing better. For a given economy, *ceteris paribus* productivity increases are always a good thing. However, it is problematic to compare productivity levels across two heterogeneous economies as a way to conclude anything about relative well-being. This is true because economic agents across the two economies are making work decisions in a completely different setting.

European governments generally provide disincentives to work long hours through progressively high income tax rates. Such governments reinforce this leisure-inducing effect through relatively large transfers to lower-income citizens. In much of Europe, a great deal of government redistributive activity encourages short and relatively productive work routines. Moreover, the *European Union* enforces an outright ceiling upon the length of a workweek. Without special dispensation, *EU* workers are limited to 48 hour workweeks.

The US government is, in general, less redistributive of income than are European countries. Thus, high-income workers are less discouraged from working long hours, and lower-income workers are less encouraged to engage in leisure (i.e., time not working). That is, a reduced level of government redistributive activity encourages longer workweeks with somewhat reduced levels of productivity. As with the European worker, the American worker is merely responding to incentives to derive the most well-being from her career. However, the American worker's particular incentives are more geared toward work than leisure. If an American worker attempts to behave like a European worker, she will not receive the same benefit from foregoing the additional work hours. These additional work hours contribute to her (and the economy's) well-being, as evidenced by the observation that both she and the firm choose to allocate them in the first place. As such, both the worker and the economy will actually be *worse off* in foregoing these marginal hours of work. There is no real economic lesson for the average American worker in this case. No matter her nationality, a worker need not be coaxed into acting in her best interest. She will work each additional hour in which the wage exceeds the opportunity cost of working. Americans and Europeans choose a different number of work hours, on average. However, we have no reason to believe that one group is choosing the "wrong" number of hours for themselves. Both sets of people are merely maximizing welfare given the unique constraints that face them.

We can consider an allegory from the wide world of sports. In the late 1960s and early 1970s, there were two major basketball leagues in the United States: the *NBA* and the *ABA*. We observed more long-range shooting in the *ABA* than in the *NBA*, where many fans found this to be an exciting element of the *ABA* game. From this, we might conclude that *NBA* players should have been shooting more long-range shots at the time. Such a facile conclusion would be missing a key point, however. *ABA* players faced stronger incentives for long-range shots in the form of the three-point line. While one could make the argument that the *NBA* should have instituted a three-point line during that period of play, we cannot conclude based on this comparison that *NBA* players or *ABA* players themselves were shooting the "wrong" number of long-range shots. As in the case of the US and European economies, such a comparison would not be *ceteris paribus* in nature.

Unemployed Presidents

US Presidents in the late stages of their first term are excusably sensitive about the use of the word "recession." As Jimmy Carter began gearing up for his reelection campaign in 1979, economist Alfred Kahn, who then served as *Chairman* of the *Council on Wage and Price Stability*, spoke candidly to the press about the possibility of an oncoming recession. In response, *White House* officials requested that Kahn refrain from further use of the word "recession" when publicly describing economic conditions of the time. Kahn nominally complied by substituting in the word "banana" each time he wanted to say "recession." In one interview during that period, Kahn said, "Between 1973 and 1975 we had the deepest banana that we had in 35 years, and yet inflation dipped only very briefly" [4]. In a later case of a President trying to coerce a rogue economist, George H.W. Bush exerted pressure upon *Federal Reserve* Chairman Alan Greenspan to undertake an aggressive style of expansionary monetary policy during the early 1990s. Bush pressured Greenspan in the wake of a mild recession that threatened, and ultimately may have doomed, his bid for reelection. Such Presidential actions are obviously a response to expectations from American voters to deliver the economic goods. However, should late-term recessions (expansions) necessarily be taken as indicative of poor (good) executive policy?

It is well known, and perhaps obvious, that US voters judge Presidents partly on the basis of their economic policy performance. However, it is not as obvious upon what grounds such judgments are made. Winston et al. [5] find that Post-WWII US Presidents "are rewarded by voters for reducing economic regulation and for increasing health, safety, and environmental regulation." Along similar lines, we can consider the bearing that macroeconomic policies and outcomes have upon Presidential elections. Since the end of WWII, eight US Presidents have sought reelection.[2] Six of the eight Presidents were successful. It is interesting, and potentially telling, to note that four such Presidents, Dwight Eisenhower, Ronald Reagan, George W. Bush, and Barack Obama, encountered the onset of a recession, as documented by the *National Bureau of Economic Research* (*NBER*), during the first half-year of their respective first terms (i.e., before any of their fiscal initiatives could conceivably affect macroeconomic conditions).[3] In each of the four cases, the recession ended before the halfway point of the term, and the President subsequently won reelection. Jimmy Carter and George H.W. Bush were the only two Presidents in the Post-WWII era who have failed to win reelection. These two Presidents encountered less fortunate timing with respect to macroeconomic conditions. Each began his first term clear of economic recession but met with the onset of a recession in the middle or late part of the term. The other two Presidents to run for reelection since WWII, Nixon and Clinton, began and ended their respective first terms clear of recession and won reelection.

Despite a paucity of data on the matter, these voting records provide some evidence that Americans associate our economic lot with the actions of the President. Voters appear to love a comeback story (i.e., a President who inherits a lemon and finishes the term with lemonade) or a case in which the President seems to keep a good thing going. However, a President who appears to lead us into the muck is unlikely to be invited back for an encore performance. Figures 3.2 and 3.3 display the timing of first-term recessions for the two Post-War Presidents who failed in their respective bids for reelection. Figures 3.4, 3.5, 3.6, and 3.7 illustrate the timing of first-term recessions for the four Presidents who succeeded in their respective bids for reelection after experiencing an early-term recession. Of course, we must remember that the two remaining reelection bids featured no first-term recession and were successful.[4] Figures 3.2, 3.3, 3.4, 3.5, 3.6, and 3.7 are depicted as follows.

From Figs. 3.2, 3.3, 3.4, 3.5, 3.6, and 3.7, it appears that the timing of a recession's onset is quite important to Presidential reelection hopes. Of course, Presidents do not control the condition of the economy at the time they begin office. By and large, however, voters apparently fail to consider that an economy's starting point plays a large role in determining economic conditions three or so years hence. Of the 713 months directly following World War II (September 1946 through December 2004), the United States spent 114 months in recession and 599 months out of recession according to the *NBER*. That is to say, our economy spent 84% of

---

[2]This number does not include Lyndon Johnson, who halted a re-election bid in the first month of the 1968 Democratic Primary race. Johnson's decision was fueled by Party turmoil and cardiovascular health problems, and his chances of re-election had he continued the bid are unclear.

[3]The NBER recessionary dating procedure takes into account not only percent changes in GDP but also unemployment rate and percentage of industries with declining employment. This procedure is commonly accepted within the economics profession.

[4]As the NBER has recognized in its recessionary dating procedure, a recession is multifaceted. Therefore, it would be incomplete to feature graphs of GDP fluctuations over time to characterize economic conditions during a Presidential term. Especially when voter preferences are considered, distributional economic issues, such as the unemployment rate, are important to consider.

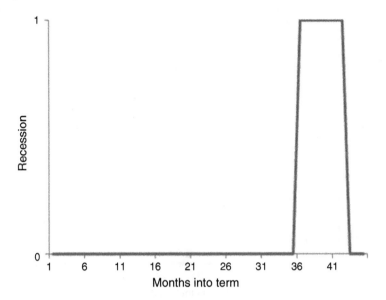

Fig. 3.2 Timing of recession during Carter's term. Source Data: *National Bureau of Economic Research*, Business Cycle Dating [6] . Note: This timeline runs from Carter's first full month of Presidency (February 1977) through the last full month of his bid for reelection (October 1980). Note also that the variable along the vertical axis takes on the value one during recessionary months and zero otherwise

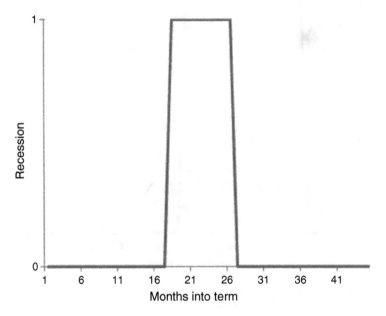

Fig. 3.3 Timing of recession during George H.W. Bush's term. Source Data: *National Bureau of Economic Research*, Business Cycle Dating [6]

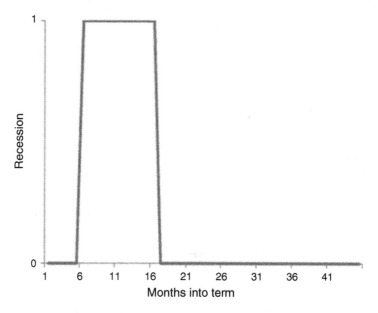

Fig. 3.4 Timing of recession during Eisenhower's first term. Source Data: *National Bureau of Economic Research*, Business Cycle Dating [6]

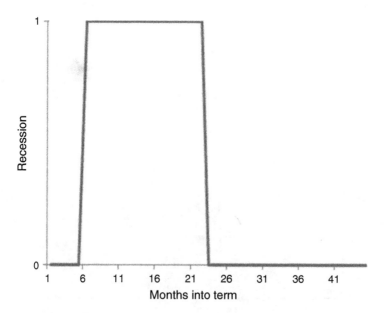

Fig. 3.5 Timing of recession during Reagan's first term. Source Data: *National Bureau of Economic Research*, Business Cycle Dating [6]

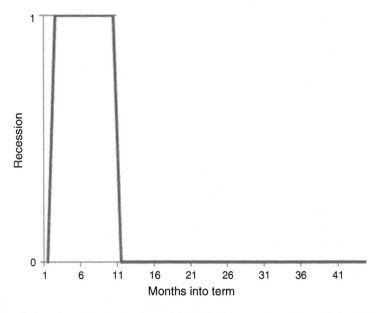

Fig. 3.6 Timing of recession during George W. Bush's first term. Source Data: *National Bureau of Economic Research*, Business Cycle Dating [6]

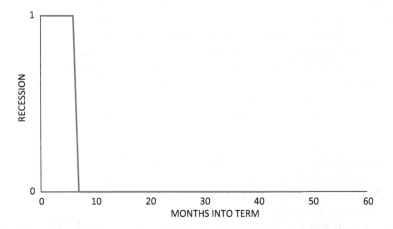

Fig. 3.7 Timing of recession during Barack Obama's first term. Source Data: *National Bureau of Economic Research*, Business Cycle Dating [6]

those months out of recession. Over that time period, the likelihood of being out of recession 3 years hence was 93% (106 of 114) for recessionary months but only 82% (493 of 599) for non-recessionary months.

Inheriting a recession or experiencing a recession in the first months of office significantly increases the likelihood that the economy will be out of recession amidst the subsequent election cycle. A test for the difference between these two percentages is statistically significant at the .0001 level, meaning that such a difference is highly unlikely to arise by chance. Further, this result accounts for all recessions in the first 57 years after World War II, regardless of Administration or timing in the electoral cycle. Therefore, the difference informs us more about the nature of short-run macroeconomic fluctuations than about the fiscal performance of an individual President. When assessing the economic performance of a first-term President, Americans may be rewarding little more than good timing.

Wage-Life Tradeoffs

We often hear that life is priceless. But is this statement true? Obviously not. If it were, we would always choose the action expected to maximize our life expectancy...At any cost. We would eat the salad instead of the burger, skip out of that really great party early to ensure sufficient sleep, and almost never set foot in a motor vehicle. In short, we would exhibit the self-control of a monk. As we know, the above choices—and the levels of self-control that they imply—do not accurately describe how humans behave. In fact, we trade away life expectancy every day for a host of other finitely valued goods. Even Bob Wiley, the lovably omni-phobic character in the movie *What about Bob?*, finally gets on the bus to see his psychiatrist at Lake Winnipesaukee. While at the lake, taking a vacation from his problems, Bob subsequently endures the risks associated with "becoming a sailor" in exchange for some quality time with his psychiatrist's son. Such behavior is not uniquely indicative of Bob Wiley, either. Consider the following real-life examples, in which people trade a little life expectancy for additional future income.

It was early 2005. Eddy Curry, who at the time was a promising young *NBA* player, had what looked to be a very good life. He was making about $3 million per year to play basketball, a game greatly simplified by his 7′0″, 295 lb. body frame. Curry was about to sign a new contract with the *Chicago Bulls* that would pay him closer to $10 million per year. Things started to get complicated for Curry, however, when he was diagnosed with arrhythmia. Before offering him a new contract, the *Bulls* demanded that he go through DNA testing to determine the risk level associated with his condition. Curry refused to do so [7] and eventually was traded to the *New York Knicks*. We can gain some insight about life valuation by examining Curry's decision to forego DNA testing.

It is obvious why the *Bulls* wanted the DNA test administered. The organization wished to determine whether Eddy Curry was a young, talented, 7′0″ basketball player with a temporary arrhythmia or a young, talented, 7′0″ basketball player with a genetic heart defect. Clearly, if the second case turned out to be true, Eddy Curry stood to lose in terms of future income. With the certainty of a genetic heart defect, not only would Curry have been less valuable to the *Bulls*, he would have been less valuable to *any NBA* team. However, Curry stood to gain from the DNA test in another sense. Over the average of all possible realities, he could have used the test information to find a level of physical activity consistent with a greater life expectancy. In choosing to forego the test, then, he essentially traded the possibility of a longer life for cash, albeit truckloads worth of cash. The possibility of a heart defect was not a trivial concern for Curry. Basketball is a cardiovascularly intensive sport that presents a substantial risk for those with heart conditions. In 1990, a star collegiate player named Hank Gathers collapsed and died of a heart attack while on the court as a result of a condition called hypertrophic cardiomyopathy.

So we've found an example of somebody trading away a little life, in expectation, for a lot of cash. But does such behavior occur in job markets more common to the average person's experience? The answer, again, is *yes*. Some construction teams build one-story structures, while others build 100-story structures. Why would a construction team with abundant opportunities to bid for 1-story projects ever take on the additional fatality risk associated with a 100-story project? [8] The answer, of course, is that 100-story projects are more lucrative. In economics jargon, these projects pay a *compensating wage differential*. Much academic work has been done in this area. The research shows that workers in many industries make life expectancy, income tradeoffs.

A study by Jeff DeSimone and Edward Schumacher [9] finds that registered nurses sometimes decide to treat more infectious patients in exchange for a salary increase. Similarly, Brandon Payne [10] shows that some police officers decide to join a slightly more dangerous unit or a slightly more violent police district in exchange for additional salary compensation. Compensating wage differentials are not specific to particular labor markets. Rather, they are found to be a general feature of labor markets. Kniesner and Leeth find [11] that US manufacturing workers are typically paid a 1% premium in exchange for working in a job of average fatality risk rather than one that features no fatality risk. For a $40,000 per year job, this estimated premium would be in the neighborhood of $400.

This does not seem like a lot of money. However, we must remember that the average job fatality rate across all occupations is still very low. In the United States, this rate is approximately 3.5 per 100,000 full-time equivalent worker years according to the *Bureau of Labor Statistics* [12]. In moving from a job with no fatality risk to one with an average fatality risk, a US manufacturing worker is paid a $400 premium in exchange for accepting a 1 in 28,571 chance of meeting with job-related fatality during that year. From this, we can roughly extrapolate that American manufacturing workers value their lives by an amount equal to approximately 28,571 · $400 or approximately $11.43 million. This estimate is in line with estimates of implied life valuation from the economics literature.

Kniesner and Leeth find similar results in the cases of Australian and Japanese workers. Their estimates indicate that individuals value life by a lot, but life is not priceless. In many ways, the "pricefulness" of life is a good thing. You probably couldn't even get a pizza delivered to your house in a world of priceless lives. As relatively safe as it is, nobody would take the delivery job. Moreover, we would have trouble with the risk of moving around and socializing in general such that pizza may not have been invented in the first place. Also, *you* might never have been invented because none of your would-be maternal ancestors would have accepted the mortality risks associated with childbirth. So you and your favorite delivery pizza are in some respects the product of life not being priceless. Of people taking big and small risks during their lives.

All of these examples and studies tell us one basic thing: *Life is generally highly valued but not priceless.* Whether or not we all wish to admit it, our actions show that no good, not even life itself, is sacred in trade. While this result may appear cold, its recognition in policy circles can have the perhaps ironic effect of saving lives. By understanding how individuals treat life itself, we enact new public safety laws that save additional lives and rid ourselves of existing ones that are misguided.

The Dynamic Injustice of Income Taxes

The typical professional athlete makes a high income for a short time period. The typical person makes a relatively low income for a long time period. Given these observations, there is some injustice in the federal income tax system. Namely, the federal income tax does not take into account whether a person's current income level is sustainable. To illustrate this point, let us assume that Mr. Athlete and Mr. Average are both 21 years old and entering the workforce. Mr. Athlete is expected to enjoy a 4-year career as a professional athlete and make $500,000 each year. Subsequent to his athletic career, assume Mr. Athlete will exit the workforce forever. Mr. Average, on the other hand, is expected to make $50,000 each year for 40 years. After those 40 years, Mr. Average will exit the workforce forever. For simplicity, let us treat each yearly income as being given in present value terms taken at the time a given worker enters the labor force. As such, both men are expected to earn $2 million over their respective lifetimes. However, Mr. Athlete expects a lumpy income stream, making all his lifetime income as a young man, while Mr. Average anticipates a smooth stream of income. Given their equal earnings behavior, does the federal income tax system treat these workers equally?

On the contrary, Mr. Average must pay about twenty percent of his lifetime income, or $400,000, in federal taxes. Mr. Athlete, given the same lifetime income level, must pay about 33% of his lifetime income, or $660,000, in federal taxes. Despite earning the same lifetime income magnitude, Mr. Athlete is punished to the tune of $260,000 merely for the lumpiness of his income stream. A "smart" federal income tax system would recognize the fleetingness of Mr. Athlete's income level and tax him at a lower rate based on his expected lifetime earnings. Whether or not fair, the purpose of progressive taxation is to draw public funds from high-income households at a greater rate than from low-income households.[5] This purpose should hold as irrelevant the intertemporal manner in which a worker receives income payments.

[5]The acute reader might point out that a tax cut for Mr. Athlete would shift his labor supply curve right and thus reduce his gross wage level. However, Mr. Athlete's net (after tax) wage would increase, along with his quantity of employment. Thus, there exists a range of tax cuts that would make Mr. Athlete better off while also reducing the level of tax injustice he faces.

Marx's Labor Theory of Value and the New Water-Diamond Paradox

When going to the beach, I usually don't enter the ocean due to an acute fear of becoming shark jerky. My friends sometimes tell me that shark attacks are very rare and that I'm more likely to die on the drive to the beach than at the jaws of a shark. The implication of this statement is that I should have no reservations about swimming in the ocean if I am willing to accept the far greater risk of driving to the beach. However, such logic takes into account only the *costs* of the respective activities and not the benefits. It is generally costlier, in terms of safety risk, to drive to the beach than to swim in the ocean once there. However, a person might derive more benefit from driving to the beach (i.e., being at the beach rather than at home) than from swimming in the ocean once there (i.e., enjoying the ocean rather than enjoying the oceanfront). You see, the quantity of beachgoers is determined by the supply of *and* demand for beach trips. Similarly, the quantity of ocean swimmers is determined by the supply of *and* demand for ocean swims. Thus, it might be perfectly consistent for a person (i.e., me) to accept the risk associated with driving to the beach but decide that the risk of actually getting into the water is too great.

This isn't the first case in which people have analyzed a market from only one side. The original water-diamond paradox asked why water, which is essential to life, costs so much less than diamonds, which are a comparative frivolity. This apparent economic paradox persisted as long as people considered only the demand side of the markets for water and diamonds. The incomplete logic driving the paradox went as follows: *Water, a necessity, is strongly demanded. Therefore, its price should be much higher, in proportion to the price of diamonds, than is observed.* However, the price of a good is not determined by demand alone. We must also consider the cost of producing the good. It is costly to bring diamonds to market. A diamond is a highly scarce mineral that must be located, mined, cut, and polished. On the other hand, it is relatively cheap in most regions of the world to bring water to market. To obtain water, a person must dig a hole, walk to the creek, or, in some cases, merely open her mouth to the sky. Thus, diamonds are priced so highly compared to water because a good's price results from the interaction of wants *and* means in the marketplace.

Karl Marx didn't quite grasp this last point. In fairness to Marx, all of us who are able to read this passage have had the benefit of twentieth and twenty-first century thought, while he didn't make it out of the nineteenth century alive. Marx believed that a good's price is determined by the quantity of labor required to produce it. Full stop. However, as the previous paragraphs have instructed us, consumer wants have a little something to do with price as well. Consider the following two goods: (1) *one well-preserved Stradivarius Violin made in the brand's "golden period" during the early 1700s*, (2) *100,000 Mendini beginner violins made in 2019*. Which bundle of goods required more labor hours to produce? In total, it required many more labor hours to make and sell 100,000 beginner *Mendinis*. On the other hand, which good would fetch a higher market price? A well-preserved *Stradivarius* made during that era can fetch \$10–\$15 million, whereas 100,000 beginner *Mendinis* would cost \$8,999,000 plus tax on *Amazon*. In this case, exceptionally high demand for golden period *Stradivarius* violins has pushed prices to astronomical levels. The level of demand owes to the heretofore unmatched sound quality of these instruments, which has caused a global bidding war. Investment funds have been started by groups hoping to pool funds toward the purchase of a *Stradivarius*. On the other hand, there are almost certainly no registered investment funds devoted to the bulk purchase of beginner *Mendinis*, which generate a sound quality that is adequate but not exceptional. In this case, factors such as labor skill, raw material quality, and demand for a rich violin sound that complements the human voice combine to dominate quantity of labor in dictating market prices. It is often the case in markets that consumer tastes and various production-side variables factor substantially in the determination of market price.

Amazon's Other Rating System

In addition to *Amazon Prime* movie ratings, *Amazon* features a broader set of customer-driven product ratings throughout its main site. Unlike *Amazon Prime* ratings, product ratings on *Amazon* do not use a collaborative-filtering approach. Rather, an *Amazon* average rating for a given product simply represents the average of all user ratings for that product. As such, all *Amazon* customers will see the same average rating for a given product, regardless of customer characteristics, purchasing history, and rating history. It may seem curious that *Amazon* rates products so differently than it rates movies. However, we should remember that most movies on *Amazon Prime* are rated extensively, whereas many *Amazon* products are not. Given that most *Amazon* products lack a large sample of ratings, it would be difficult to identify clusters of customer types for given products throughout the site. That is, product ratings are generally not as *thick* as user ratings.

There are ways to improve the company's product rating model, however. As *Amazon* is involved in so many technologies, some of these improvements can be achieved simply by combining two or more of the company's present competencies. For example, *Amazon* invests heavily in artificial intelligence systems. At the same time, *Amazon* product ratings commonly feature an AI-soluble error. Namely, users often rate the *seller* of a product within the product rating survey. This is often obvious, as such erroneous product ratings often feature comments to the effect that the product came late or, in the case of a used item, damaged. Such issues of distribution or used item condition should not reflect poorly upon the underlying product. Product ratings are meant to tell prospective buyers about the quality of the product itself apart from any confounding factors. Seller ratings posing as product ratings can greatly distort user evaluations of a product's underlying quality. Of course, *Amazon* has the resources to clean up many of these misplaced ratings. Whenever such a rating is accompanied by text that evaluates distribution or used

Table 3.1 Product ratings and reviews for a fictional book titled *Cake Pops for the Battered Soul*

User	Product rating	User review comments	# People finding review helpful
Khan	5	Well-written book, concise, informative	7
Anderson	3	Nice cover!!!	3
Miles	1	Bought used copy from jjredickno1. Didn't come on time. Totally bummed.	0

Fictional author-constructed example

item condition, a natural language processing based artificial intelligence algorithm could simply purge the rating or redirect it to the appropriate seller rating. Such a fix would allow *Amazon* to attain more accurate product ratings without any need for new research and development investments. The remedy would be based upon technologies they currently possess.

Amazon does not simply ask customers to rate products. The company also allows customers and prospective customers of a product to rate each product rating in terms of helpfulness. That's not a typo. *Amazon* provides a mechanism for ratings of ratings on each product page by asking whether each given rating was helpful or not. From this rating of ratings, *Amazon* possesses and even lists data on the number or proportion of responding users who found a particular product rating helpful. By combining its product rating and rating helpfulness data, *Amazon* could likely improve upon its average product ratings by weighting each product rating according to its degree of helpfulness and then calculating a weighted-average product rating. This weighted-average measure would be expected to provide more predictive estimates of future customer experiences, as it would tend to "down-weight" reviews that are unhelpful or misplaced and "up-weight" reviews that are well supported.

It may be that some readers do not understand the distinction between a simple average and a weighted average. If so, the previous paragraph was a load of gibberish. Let us go through a simple example to clarify the distinction between simple and weighted averaging. Specifically, imagine a book titled <u>Cake Pops for the Battered Soul</u> that has three reviews. Rating and review information for the book are provided in Table 3.1.

The simple-average user rating for this book is simply $(5 + 3 + 1) / 3 = 3$ stars. However, this average is pulled down substantially by the review of Miles. The effect of Miles' review is unfortunate, as he has clearly written a seller review in the product review survey space. We know this from the comments, and we also know that other users have picked up on this. No users have marked Miles' review as helpful. Let us now calculate the average product rating over all reviews weighted by the number of other users finding each review helpful. To do so, we can simply multiply each rating by its total number of helpfulness attributions and then divide by the total number of helpfulness attributions overall. Following this method, the weighted-average rating for this book is $[(5 \cdot 7) + (3 \cdot 3) + (1 \cdot 0)]/10 = 44/10 = 4.4$ stars. <u>Cake Pops for the Battered Soul</u> has jumped from a 3-star simple-average rating to a 4.4-star weighted-average rating. We expect that the new rating better reflects the book's true quality for users moving forward, as it makes use of two complementary layers of user survey data. We also note that the weighted average product rating has weeded out a misplaced seller rating in this case. As no users found the misplaced rating helpful, it did not reflect at all in the final weighted average rating. This weeding out was achieved not by *AI* in this case but, rather, by good old-fashioned *I*. By not recognizing the review as helpful, users effectively cleaned the data that was applied to the weighted-average rating calculation.

A Case for the Celebrity Treatment

Crime occurs because it carries a sufficient benefit to the successful perpetrator. This is not to say that lawless activity is costless. Indeed, punishment of crime is meant to alter the calculus of the criminal by increasing her expected cost. A successful punishment causes the criminal to realize that she will achieve more well-being through lawfulness than through crime. In order to meet this goal of future deterrence, a punishment must become more severe as the potential benefits from crime rise or as the likelihood of apprehension falls. Punishment is commonly achieved through incarceration but can also be achieved by directly fining the apprehended criminal. A monetary fine equal to the amount the perpetrator would pay to avoid incarceration is equally deterrent as incarceration itself. Rather than meant to stir controversy, the prior statement is something close to a tautology (especially if read carefully). At the same time, this incarceration-equivalent fine does not carry the administrative cost of incarceration. Thus, fines can be an efficient form of punishment for a broad class of crimes. In the case of many celebrity wrongdoings, there is an added incentive for society to forego administering punishment through incarceration, as was illustrated in the case of Michael Vick.

In 2008 and 2009, former star *NFL* quarterback Michael Vick spent 18 months in prison after pleading guilty to charges that he staged several illegal dog fights. While incarcerated, Vick did not play in the *NFL*. At the time, however, he was extremely valuable to society as a quarterback. Given his unique talents on the field, Vick created an entertainment service worth many millions of dollars per year as an active player such that the *NFL* lost substantial entertainment value, in expectation, during his absence. This decreased value can be approximated as the difference between Vick's yearly *NFL* salary and the salary of the marginal or replacement level *NFL* quarterback in 2008–2009. Factoring in expected performance bonuses, this difference in salaries was likely equal to more than $10,000,000 per season that Vick missed.

As an inmate, Vick produced nothing of substantial value. Incarcerating Michael Vick was apparently beneficial to society in the sense that it deterred him and others from committing similar acts in the future. At the same time, it was quite costly in that it denied him the opportunity to play football and others the opportunity to watch him play. An optimally priced fine, perhaps paired with heavy community service, surveillance, and off-season house arrest, would have provided a similar deterrence effect while allowing Vick to create economic value on the field. Moreover, proceeds from the fine could have been used toward improvements in law enforcement and animal rights. In the case of highly valued individuals such as Vick, therefore, punishment by incarceration carries a high opportunity cost to society. This is not to suggest that these individuals should receive a lesser penalty. Indeed, there is no unequivocal justification for penalizing them less than other members of society for a particular crime. However, the way in which the penalty is levied has a strong bearing upon our collective social welfare.

Countless readers might suggest that, despite his skill level, Michael Vick was no longer of value as a football player after the fallout from his transgressions. While certainly less popular than before, Vick remained of considerable value as a football player. As difficult as it might be for some people to believe, many sports fans seem to be adept in separating the player from the person. *ESPN SportsCenter* has never done a segment on whether 22-year *NBA* veteran Vince Carter does the dishes for his wife regularly while at home because the typical *NBA* fan is not passionate about Vince Carter's at-home persona. Similarly, if a star *NFL* quarterback spent his weekdays taking lollipops from the mouths of local children, this hobby would not affect the enjoyment that many fans derive from watching him complete passes on Sunday. If this weekday hobby helps him feel more dominant on the football field, some of his more ardent fans might even think, *"The more power to him."* While there are undoubtedly some readers who loathe my current line of reasoning (I hope those readers can separate the author from the person), there are countless such sports fans. A 2014 survey of the American public from globalsportsdevelopment. org found that two-thirds of respondents "admit they sometimes suspect the athletes they are cheering on might be using PEDs" [13]. And yet they cheer for them. Unless these fans have even stronger suspicions about the athletes they're not cheering for, it seems as though suspected performance-enhancing drug use does not constitute a deal breaker for most American sports fans.

Quarterback Injuries as a Negative Externality of Hard Defensive Play

When Tom Brady sustained a season-ending knee injury to begin the 2008 *NFL* season, my interest in football, as a "fan of quality play," fell immediately. Suddenly there was unlikely to be any of the record-setting magic that *NFL* viewers witnessed in 2007. The injury, which occurred when *Kansas City Chiefs* safety Bernard Pollard rendered a knee-level tackle on Brady, motivated me to think about how the *NFL* can curb injuries to (star) quarterbacks. There are already various measures in place to protect the *NFL* quarterback. For example, late hits on the quarterback are closely watched and heavily penalized. However, defensive players are in no way penalized for executing knee-level, or otherwise dangerous, tackles on the quarterback. As defensive players bear none of the costs associated with quarterback injury, they will hit around the knee even when the benefit of such an action, in terms of the play's statistical value, is exceeded by its total social cost to fans, the quarterback's team, and the quarterback himself. If a defensive player were liable for all expected costs associated with a knee-level tackle, he would be much more hesitant to make such a play.

Assume that *Defensive End A* injures *Quarterback B*, a star quarterback, by tackling him at the knee. Further, assume that the tackle effectively rushes the quarterback's throw. As the defensive end's performance evaluation greatly depends upon how many times he sacks or rushes the quarterback, the play is worth, say, $25,000 to the defensive end in terms of gains to expected future income. However, imagine that the play costs the quarterback the rest of the season. Although the quarterback is paid for the remainder of the year (i.e., there are no immediate monetary costs borne by the quarterback), fans of quality play such as myself face an expected loss of entertainment value in Quarterback B's absence. Further, the team and league potentially lose a valuable asset. Of course, there can be offsetting gains from a quarterback injury. Tom Brady's abilities were "discovered" in the wake of an injury to then-teammate Drew Bledsoe. Brady's own absence allowed for the "discovery" of Matt Cassell, who subsequently signed a 6-year, $63 million contract with the *Kansas City Chiefs*. In expectation, however, a major injury to a star quarterback in his prime drives the League's quality-of-play downward.

If each fan of quality play receives $1 less value in watching a season unfold in the absence of Quarterback B and if there are a million such NFL fans, the immediate cost borne by fans of a season-ending injury to Quarterback B would be $1 million. Further, this loss of value to such fans would continue into future seasons if Quarterback B does not fully recover, a consideration that will affect the quarterback's future contract negotiations. Let us assume that fans face a *future* quality-of-play loss of $1 million, in expectation, due to Quarterback B's present injury. In sum, Defensive End A has imposed a multimillion dollar cost on society to take an action that netted him $25,000. He is willing to take such an action because he bears none of the expected cost associated with the action. In other words, hard and somewhat reckless play on the part of a defensive end incurs a negative externality upon others. If the expected cost associated with a knee-level hit were transferred to the tackling defensive player, it is expected that the latter player would magically begin aiming higher when making contact with the quarterback. Of course, such fines could apply in the case of (injurious) ankle-level tackles, tackles in which the quarterback is slammed to the ground, and "excessive force" blindside tackles.

In the face of such stiff fines, it is further likely that defensive players would collectively insure themselves. That is to say, a defensive player who anticipates $5 million in career earnings would not look kindly upon the possibility of a $5 million fine. Pooled insurance funds would likely be used to pay injured quarterbacks when a defensive player is deemed to be at fault for a quarterback's knee injury. Even with full insurance, defensive players facing quarterback injury fines would be more careful in how they tackle the quarterback. In the same way that fully insured drivers exercise some care to suppress future insurance premiums, insured defensive players would exercise additional care when tackling a quarterback in the presence of injury fines.

One might wonder why this essay makes such a fuss about the quarterback position in particular. After all, aren't quarterbacks pampered enough? Despite being commonly reputed as soft prima donnas, quarterbacks absorb many hits over the course of a season. Due to the demands of the position, moreover, quarterbacks are particularly vulnerable to hits they did not see coming. On each play, the onus is on the quarterback. He must make several split-second reads following a snap and subsequently synthesize this information to decide how best to advance the ball. While in the pocket, moreover, quarterbacks usually maintain a body position amenable to throwing the ball. This pose forces the quarterback to turn his back to one side of the defense, the blind side, such that he is vulnerable to surprise hits from that side. Indeed, blindside hits can end careers, as occurred in the case of Joe Theismann.[6] The quarterback position is one of the most demanding and vulnerable positions in all of sports but also one of the most exciting to watch. For these reasons, quarterbacks are almost singular in their need for a set of game rules that serves to protect their health.

An East Coast/West Coast Thing

For decades, Jay Leno and David Letterman taped their respective late-night shows on different coasts. Today, Jimmy Kimmel and Conan O'Brien tape their respective shows on a different coast than Jimmy Fallon and Steven Colbert. There appears to be a spacing rule for the locations of late shows. When Conan briefly took over *The Tonight Show* in 2009, he became direct competitor to David Letterman, who taped his show in New York City. Before beginning *The Tonight Show*, Conan decided to leave New York City, where he had taped for years, to Letterman and move to the left coast. In the primetime slot, it appears that even NYC was not big enough for the both of them.

Elementary game theory can shed light upon the location decisions of late show competitors over the years. In the simple model of late show location to follow, let us consider a time in which there were two dominant late show hosts, Leno and Letterman. We further assume that ratings for each show are directly related to the average star power of the show's guests. Thus, each host locates where his show can maximize guest star power given the locating decision of the other host. This objective effectively narrows each show's set of possible locations to two cities: Los Angeles and New York City. Stars tend to cluster in these two spots and much less so in places like Cleveland, Ohio. Why don't late-night hosts simply share either L.A. or New York? If both Leno and Letterman choose Los Angeles, they will compete fiercely for the strong but still limited talent supply there, while largely neglecting the non-negligible talent supply available in New York City. A similar outcome will obtain if both locate in New York City. However, if Leno locates in one city and Letterman in another, each will hold something like monopsony power

---

[6]Footage of the play is not for the faint of heart.

Table 3.2 Where should I put my show?

Leno		Letterman	
		LA	NY
	LA	(9.5,9.5)	(12,10)
	NY	(10,12)	(8.5,8.5)

Source: Fictional author-constructed example

over his city's celebrities. If a New York City-based celebrity wishes to promote a new Broadway show, it will be costlier, in terms of time, for her to jet off to Los Angeles. Similarly, if a Los Angeles-based celebrity wishes to promote her new movie, she is more apt to do so in Los Angeles.

A game matrix can summarize each show's payoffs for this example, in terms of average show rating, for each combination of show locations. In doing so, the matrix helps us understand what each person will choose to do conditional upon his beliefs about what the other person will do. The first number in each cell of the matrix always refers to Leno's payoff for a given outcome, and the second number always represents Letterman's payoff, where the numbers provided are hypothetical (Table 3.2).

There are a few notable characteristics of the hypothetical payoff structure considered above. First, we assume that Los Angeles has the lion's share of talent. If the two hosts locate in different cities, the one in Los Angeles will always come away with better ratings. Second, it is assumed that the two hosts are equally popular in a given location. In other words, payoffs are strictly determined by relative locations in the example. A Nash equilibrium outcome is an outcome of the game whereby each person's decision constitutes a best response to the other person's action. This game features two Nash equilibria—one in which Leno is in Los Angeles and Letterman is in New York City and another in which Leno is in New York City and Letterman is in Los Angeles. That is to say, given a belief that his competitor will locate in one city, each host is best off to locate in the other. This is a type of *anti-coordination game*. In the game, neither party wishes to split a single talent market.

If there are two possible Nash equilibria in this locating game, how did Leno end up in Los Angeles and Letterman in New York City? It may be random chance, within the game considered, that Leno got Los Angeles. It may be that Leno places a higher premium on warm weather than does Letterman, a circumstance that would alter each host's private payoffs. On the other hand, the nature of the locating game between these hosts might be more complex than that which we considered above. Recognizing that *The Tonight Show* was established in Los Angeles long before Letterman's *Late Show* existed, we could consider the locating game as a sequential move game in which Leno enjoyed a first-mover advantage in Los Angeles such that Letterman was dissuaded from entering that market. Also, we might consider asymmetries in the popularity of the two hosts. Whatever complexities we add to the game, our original framework suggests that no town is big enough for both Leno and Letterman.

A Populist Commissioning

At my local music store, I can buy a new copy of the album *Love and Theft* for $15. Alternatively, I can buy a used copy of the same album for $8. Given that the used copy is refundable in the case of a defect, the latter product seems to constitute a nearly perfect substitute for the former. Considering their similarities, why would anyone faced with such a choice buy the new album copy? Let us assume that I value owning a copy of *Love and Theft* by $25. My consumer surplus (i.e., value above price) in buying a new copy, therefore, is $10 ($25 minus $15). On the other hand, my consumer surplus in buying a used copy of the same album is $17 ($25 minus $8) less the expected inconvenience cost of making an exchange should the used product malfunction. Liberally assuming that I am inconvenienced to the tune of $10 should the used product malfunction and that the product malfunctions 10% of the time, I am still left with $16 of expected consumer surplus ($17 minus $1) when buying a used copy of the album.

As $16 is considerably greater than $10, it seems that a new album should never be purchased in the presence of a used album if reality is anything like the present example. However, any music store employee can tell you that such a choice is made quite often. The very fact that prices for the two items are so different implies that many consumers are unlikely to view them as nearly perfect substitutes. Rather, consumers are getting something, besides the convenience of knowing that the CD will work properly, in purchasing a new copy of the album. Unlike used album purchases, new album purchases allow the consumer to effectively vote, with dollars, on the future state of music. If a consumer wants to encourage a particular musician to continue making a particular type of music, spending dollars on new copies of her work is as good a way as any. Dollars spent on new albums can signal to the musician and her record label that a work is appealing. Further, such expenditures allow the sufficiently well-received musician to focus on music as a vocation in the long run. In buying a used copy of the same album, the consumer is not transferring dollars to the musician in support of her work.[7]

Indeed, contributing to new record sales is the antecedent of an age-old European practice in which members of a society's nobility commissioned work from a particular composer or artist. The act of commissioning artistic work rewards a talented artist, while ensuring that she continues to work and entertain. Today, buying a new copy of a musician's CD might be thought of as a populist form of commissioning. One should not underestimate the value many music consumers hold for artistic voting rights. Several friends I've known will purchase music that can easily be pirated. These friends do so on the grounds that they wish to help keep a favorite musician in business. Such consumers see the dynamic efficiency in paying for their wares, which means that less evolved, myopic people such as myself can walk into a music store and pay a lot less for a used album.

---

[7]In the sense that new CD prices partly reflect a buyer's ability to resell the CD, one might argue that the used purchase market does have an influence upon musician support.

References

1. Child Trends. (2020). Youth employment. Retrieved from https://www.childtrends.org/indicators/youth-employment
2. Siegfried, J. J., & Stock, W. A. (2004). The market for new Ph. D. economists in 2002. *American Economic Review, 94*(2), 272–285.
3. Jacobson, J. (2002). The competition for economics Ph.D.s. *The Chronicle of Higher Education.*. Retrieved from www.chronicle.com/article/The-Competition-for-Economics/46125
4. Washington Post. Yes, we'll have no banana. Retrieved from: https://www.washingtonpost.com/archive/opinions/1978/12/03/yes-well-have-no-banana/ff66f487-7e1a-4579-a199-5750373405c9/
5. Winston, C., Crandall, R. W., Niskanen, W. A., & Klevorick, A. (1994). Explaining regulatory policy. *Brookings Papers on Economic Activity. Microeconomics, 1994*, 1–49.
6. Hall, R., Feldstein, M., Frankel, J., Gordon, R., Romer, C., Romer, D., & Zarnowitz, V. (2003). The NBER's business-cycle dating procedure. *Business Cycle Dating Committee, National Bureau of Economic Research.*
7. Associated Press. (2005). Bulls deal Curry after DNA test refusal. Retrieved from: https://www.espn.com/nba/news/story?id=2180298
8. Cosgrove, B. (2013). Mystery in the sky: A legendary photo (slowly) gives up its secrets. Retrieved from: https://time.com/3449718/mystery-in-the-sky-a-legendary-photo-slowly-gives-up-its-secrets/
9. DeSimone, J., & Schumacher, E. J. (2004). Compensating wage differentials and AIDS risk. No. w10861. *National Bureau of Economic Research.*
10. Payne, B. (2002). Estimating the risk premium of law enforcement officers. Retrieved from https://pdfs.semanticscholar.org/e32e/00a9a1f681ecbfd11dad698f7daafde0c489.pdf
11. Kniesner, T. J., & Leeth, J. D. (1991). Compensating wage differentials for fatal injury risk in Australia, Japan, and the United States. *Journal of Risk and Uncertainty, 4*(1), 75–90.
12. Bureau of Labor Statistics. (2019). National census of fatal occupational injuries in 2018. Retrieved from https://www.bls.gov/news.release/pdf/cfoi.pdf
13. Global Sports Development. (2014). Doping survey reveals public opinion. Retrieved from http://globalsportsdevelopment.org/2014/01/28/doping-survey-reveals-public-opinion/

Information and Misinformation

4

> *The quality of our decisions is based in part on the amount and quality of relevant information in our possession. This chapter expounds upon this simple point across various settings.*

Heterogeneous Preferences and Movie Reviews

I always check a few movie reviews before heading to the theater. This ritual helps lower the risk that I spend $12 and two hours on a movie that infuriates my cinematic sensibilities. Indeed, many moviegoers utilize critic ratings to predict how well they will like a particular film. Movie reviews are glaringly imperfect, however, in that the reviewing film critic has a unique set of movie tastes. No matter how "expert" a critic is supposed to be, she knows nothing of your or my movie preferences. Therefore, she cannot provide a good prediction of a movie's quality for every moviegoer. Indeed, reviews have caused me to watch movies that I found to be boring, as well as to skip movies (during their theater release period) that I later found to be quite good. Hence, it is risky to trust a single film critic to make your movie-viewing choices for you. For the typical moviegoer, a better approach toward estimating a movie's quality is to calculate its average rating across a number of critics. Sites like *Rotten Tomatoes* and *Metacritic* feature such average critic ratings, as well as the average crowd ratings for each film. *Average* movie ratings are a dependable guide for the typical cinematic consumer in that they iron out any extreme or unrepresentative views toward a particular movie.

However, even *average* ratings can be misleading if one's movie tastes are atypical in some way. In such a case, a moviegoer's most dependable "critic" would be a person or group of people with similarly atypical movie tastes. That said, the most relevant movie rating platform would likely take the form of a website that asks users to rate movies that they have seen. The site could use this information to place users of similar taste (i.e., people who rated previously viewed movies in the

S. Sanders, *The Economic Reason*, https://doi.org/10.1007/978-3-030-56043-0_4

most similar manner) into one of several sub-groupings or clusters. When deciding whether to see a new movie, a site user could obtain an average rating from those members of her group who have already seen and rated the movie of interest. In spirit, such a site would seek to provide a user with movie ratings from her future self. In order to provide more timely ratings, the site could even identify *critics* who are most similar to a particular user based on that user's previous ratings. In this manner, pre-release critic ratings could be used to provide an early prediction of how the user would rate a movie.

But how might a ratings website deem one user as more similar to me or you, in terms of movie taste, than another? If users were to rate movies on a scale of zero to four, the similarity of two viewers could be assessed according to the average difference in their respective ratings across all commonly rated movies. For example, imagine that Marv, Johnny, and I commonly rate one movie, *Daddy Day Care*. I give the film 1.5 stars, while Marv rates it as a one-star film, and Johnny, who really relates to the movie's zany take on fatherhood, gives the movie four stars. My rating is closer to Marv's (a half star away) than to Johnny's (two-and-a-half stars away), meaning that the website would group me with Marv before Johnny. If Marv, Johnny, and I rate twenty films, with my average difference from Marv being 0.3 stars and my average difference from Johnny being 1.2 stars, the site would, again, give Marv priority when selecting my group.

It is interesting to note, however, that a dissimilar user can be just as important as a similar user when predicting how a given person will like a movie. If Marv and I have identical movie tastes, while Johnny and I have opposite movie tastes, then Johnny's dislike of a movie is just as strong a signal that I will like the movie as is Marv's appreciation of the same movie. For more than a decade, *Netflix* has predicted user ratings for unseen films based on the experience of "like users" who have seen the movie. They do so using a modeling approach called *collaborative filtering*. *Amazon Prime* subsequently adopted a similar collaborative filtering approach. While this modeling approach is both cutting edge and useful within the environment it serves, the information rendered is ultimately limited by design. That is, each service generates predictive ratings only for movies that are "in system" (i.e., movies that their service provides). As such, their ratings do not pertain to most newly released films or to older movies that are not available in their respective systems. As mentioned previously, a website could certainly go further than *Netflix* or *Amazon Prime* if it were to a) predict a user's rating of new cinema movies based on pre-release critic ratings or b) host collaboratively-filtered ratings for a more general set of movies. Of course, such a site would have to be subscription based. One cannot have collaboratively filtered ratings unless people identify themselves through personal accounts.

The concept of predictions that are based on prior peer review can be adapted to many settings. For example, my global positioning satellite application can tell me how to get to a particular restaurant in Toronto. It can even tell me the average customer rating of that restaurant. However, a revolutionary GPS feature would ask users to rate their experience after visiting a particular restaurant or attraction but not stop there. Upon clustering users of similar taste, a GPS application could offer predictions as to which restaurants or activities a user who is new to a particular part of town might like best. The system would, again, do so based on the reviews of *like* or *unlike* users who have already visited and reviewed aspects of that part of the town. Such a program feature could allow for predictions so specific as to guide one's choice of entrée at a new restaurant. By finding our preference peers and preference opposites, a new experience need not be such a hit-and-miss proposition. Of course, such a potential feature is not ideally suited to those who prefer adventure.

When Probabilities Are Not Enough

In a *Slate* article titled "Vote!" [1], Jordan Ellenberg encourages Americans to do just that. It is conceivable, he argues, that a single vote will sway an election. Thus, the expected benefit of voting, in terms of bringing about a preferred outcome, is potentially significant. Ellenberg reasons that the odds of breaking a tie in a national election could be as great as 1 in 300,000. Compared to the odds of winning the *Powerball* (1 in 120,000,000), such a probability is enormous. If people derive benefit from playing the *Powerball*, therefore, they should certainly be motivated by the potential benefits of voting.

However, Ellenberg's comparison is incomplete. The expected benefit of an action is not determined solely by the likelihood that it will meet with a good outcome. Though the odds of winning the *Powerball* are low, the value of winning this lottery contest can be in excess of $100,000,000. Show me the individual who values Candidate A winning office, as opposed to Candidate B, by anything close to $100,000,000? To consider your own valuation for a Presidential election outcome, try the following thought experiment:

> If you won a $100,000,000 Powerball Jackpot, what proportion of your winnings would you give to a Political Action Committee in support of a particular Presidential candidate?

In general, observed contributions to candidate-specific *Political Action Committees* suggest that most of us do not value the outcome of a Presidential election by any substantial amount. Thus, a person might be completely justified in always playing the *Powerball* lottery but never voting.

Ellenberg's comparison is problematic for other reasons. Most people visit the gas station and watch the news on a regular basis. In this sense, the total *cost* of entering the *Powerball* lottery might be far less than the total cost of voting. To vote, one must (a) register, (b) drive to the polling location, and (c) potentially wait in line during work or dinner hours. Thus, Ellenberg not only ignores the value of the respective benefits associated with voting and playing the lottery, he also ignores the respective costs of these activities. That is to say, Ellenberg incompletely examines one side of the markets for voting and lottery-playing to form a judgment about voting incidence.

Even if one does gain more from voting, in terms of benefits net of costs, than from playing the *Powerball* lottery, this does not mean that people need further encouragement to vote. After all, more people vote in a typical national election than play the *Powerball* lottery during a typical week. Given that his column is titled "Do the Math," Ellenberg may have neglected the boundaries of his underlying framework in attempting to explain voting behavior. Sometimes to do the math is only one small step toward doing the economics.

The (Information) Economics of March Madness

Economics addresses the allocation of scarce resources among competing ends. In certain cases, such allocative processes are quite straightforward. For instance, if Matilda's favorite activity in life is to walk to a hilltop and watch the sunset, then she will tend to center her evenings around this event. However, some allocative processes are highly complex. Most sports leagues, for example, end the season with a tournament featuring those teams deemed to be the best. However, if a league is sufficiently large to preclude direct competition between each pair of teams, as in the case of *NCAA Division I Men's Basketball*, it can be quite difficult to accurately rank teams. A *Division I Men's Basketball* season consists of thousands of games played by hundreds of different school teams. Given geographic constraints and the large number of major colleges in the United States, no single team can hope to play even fifteen percent of all Division I teams during the course of a season. Therefore, it is difficult to use the information of a college basketball season to comprehensively rank teams.

Given the complexity of rating such a large number of teams, the *NCAA*-developed *Ratings Percentage Index* (*RPI*) was for decades the prominent measure of team ability level in *NCAA Division I Men's Basketball*. The *RPI* uses information from a team's revealed performance to numerically describe that team's ability. From 1981 through 2018, the *RPI* aided the *NCAA Division I Men's Basketball Committee* in selecting and seeding *NCAA* tournament teams. Thus, the measure held long-standing value for its ability to comprehensively and unambiguously assess college basketball teams. The *RPI* is still featured prominently on sites like *CBS Sports* [2]. It also continues to be used by several high school sports leagues to rate teams [3]. Given its widespread use over the years, it is potentially valuable to understand the methodology of the *RPI* and to determine any biases it might contain.

The *RPI* value for a given team is essentially a weighted average of that team's winning proportion (.25), the average winning proportion of the team's opponents (.5), and the average winning proportion of the team's opponents' opponents (.25).[1] By measuring the latter two winning proportions, the *RPI* is commonly believed to control for a team's strength of schedule. Noted college basketball analyst Jerry

---

[1]This is the basic formula. Some leagues adjust the RPI formula such that away wins are more valuable than home wins.

Palm states, "(The *RPI*) is a measure of strength of schedule and how a team does against that schedule."

However, the measure is problematic in that a large proportion of *Division I* college basketball games take place between conference opponents. Conferences are commonly recognized as ability-clustered collections of teams within college basketball. Therefore, a team from the mighty *Atlantic Coast Conference* requires more ability to win 60% of its games than does a team from the lowly *Northeast Conference*. The same reality holds for most of the former team's opponents and opponents' opponents. Therefore, it appears that the *RPI* is giving major conference teams, which are strongly tested in the conference season, a hard time compared to teams from lower conferences.[2] Analyst Jon Scott addresses this point [4],

> The major reason the RPI is a poor model for determining team strength is because it is too simplistic to reliably differentiate teams and relies completely on the assumption that winning percentage is a valid indicator of how strong a team is. Comparing only the won-loss percentage of the last place team of a power conference with the won-loss percentage of a low-level conference champion, with no regard for the schedule each school actually played, one would be completely misled as to which team was the stronger.

To understand the absurdity of the *RPI* formula's structure, let us consider an equally absurd example. Imagine that I challenge my grandmother to a boxing match to avenge the whipping she gave me as a 9-year-old. She and I go to Las Vegas to mix it up in front of a live audience, and I outpoint her narrowly in twelve rounds. On the undercard, Floyd Mayweather defeats Manny Pacquiao. Considering only those two matches, the *RPI* would conclude that I am as good a boxer as Mayweather and that my grandmother is as good as Pacquiao. Let us assume that I go on to defeat all of my relatives (except for my dad, who I shamelessly dodge). Meanwhile, Mayweather continues to bully the best middleweights in the world. The *RPI* would continue to rank me in the neighborhood of Mayweather and better than many of his opponents, regardless of whether I ever fight Mayweather or any boxer that he has fought. Of course, the reality is that Floyd Mayweather could make me forget my address with a single punch.

The main point is that the *RPI* does not properly control for a team's strength of schedule. Thus, we might expect a mid-major conference team that is strong relative to its conference opponents to sometimes achieve a higher *RPI* than a major conference team that is mediocre or weak relative to its conference opponents, *even when the major conference team has more ability and has exhibited this in games against strong mid-major conference teams*. For years, the possibility of a positive "mid-major bias" in the *RPI* carried important and pivotal implications with respect to *NCAA Tournament* selection and seeding, a process for which the *RPI* was

---

[2]There are two conference tiers in college basketball that are relevant to the NCAA tournament discussion- major (i.e., first tier) and mid-major (i.e., second tier). Though the distinction is unofficial, it is commonly accepted that six of the 31 Division I conferences are major conferences, while the remainder are mid-majors or low-majors. Despite their relative scarcity, major conference teams have earned 97 of 108 NCAA Tournament Final Four spots during the RPI era.

created by the *NCAA*. Though mid-major teams were slightly outnumbered in the *NCAA Tournament* during the RPI era, the *NCAA Division I Men's Basketball Tournament Selection Committee* may have been overstocking the *Tournament* with relatively strong mid-major conference teams due to a measurement system that over-estimates such teams.

Of course, the *NCAA Tournament Selection Committee* may have relied on the *RPI* for all those years not for its fairness but for its ability to draw more potential "Cinderella teams" into the *Tournament*. After all, an *NCAA Tournament* with high Cinderella potential is quite often a more interesting *NCAA Tournament*.

Assessing the Quality of Tournament Selection

Several years back, the first day of the *NCAA Men's Basketball Tournament* passed without a significant upset. In response, a sports radio anchor that I happened to be listening to commented that this outcome was a testament to the *NCAA Tournament Selection Committee*'s ability to select and seed teams. However, it is not necessarily true that a dearth of upsets indicates accurate selection. For example, we would expect fewer first-round upsets if the *Selection Committee* chose the 32 *worst* teams in *NCAA Division I Men's Basketball* to inhabit the unfavored seeds. While this selection methodology would make seedings more predictive of outcome, it would not be at all accurate in identifying the 33rd through 64th best college teams in the country. At the end of the day, the accuracy of the *Selection Committee*'s decision-making depends upon its ability to measure team quality and, as mentioned previously, on its beliefs concerning what set of teams will generate the most *Tournament* viewers. It is difficult to judge the *Committee*'s true performance because we will never know how teams that were not selected would have done in the *Tournament*. However, we can at least say that the best selection process, in expectation, is one that identifies worthy teams based on an efficient use of *all* information generated from regular season play.

Your Enemies' Enemy

The relatively uncelebrated sport of cross country running also measures team performance in an inconsistent manner. Within a given meet, a cross country team's score is calculated by summing the places of its top five runners, where a team consists of five to seven runners. Teams are then ranked in ascending order (i.e., from the lowest team score) to form a meet outcome. This ranking methodology is flawed in the sense that the relative ranking of two teams can depend upon the performance of other teams in the meet, as was shown by Thomas Hammond in an article published in the journal *Public Choice* [5]. Let us assume that Team A defeats Team B in a two-way meet. Holding team performances constant, this does not imply that Team A will defeat Team B if Team C also participates in the meet. That is to say, the scoring methodology utilized in cross country fails to rank two given team outcomes in a manner that is independent of whether or not a third team is present in the race. In social choice theory, it would be said that rank sum scoring is not *Independent of Irrelevant Alternatives.*

We can illustrate this inconsistency in an example. Imagine three teams—A, B, and C. Each team consists of five runners. Team A runners always finish a 5-km race in the following respective times: 16:00, 16:20, 16:40, 17:00, 17:20. Team B consists of a set of quintuplets, each of whom finishes every meet in 16:41 (within a decipherable fraction of a second of one another to allow for unique individual placements in the race outcome). Team C consists of another set of quintuplets, each of whom finishes every meet in 16:59. Whenever the three teams face one another in a three-way meet, then, the finishing sequence appears as follows:

$$< A , A , A , B , B , B , B , B , C , C , C , C , C , A , A >$$

In this three-way meet, Team A places runners in finishing positions 1, 2, 3, 14, and 15 and therefore scores $1 + 2 + 3 + 14 + 15 = 35$ points, while Teams B and C score $(4 + 5 + 6 + 7 + 8 =) 30$ and $(9 + 10 + 11 + 12 + 13 =) 55$ points, respectively. That is, Team B wins the three-way meet. However, when Teams A and B face in a two-way meet that features the same individual performances, the finishing sequence contracts to:

$$< A , A , A , B , B , B , B , B , A , A >$$

In this dual team race, Team A wins by a score of 25–30. Why does Team A win in the two-way meet but not in the three-way competition? Team A has relatively weak fourth and fifth runners, which hurts the team greatly in the three-way meet but not as much in the two-way meet. In a two-way meet, fewer opposing runners can finish before Team A's fourth and fifth runners. For example, the "cost" of Team A's fourth runner not running 5 seconds faster is five places in a three-way meet but zero places in a two-way meet with Team B. It is this variation in costs that leads to inconsistent team rankings across meet in cross country.

Hammond points out that the sport of cross country can fix this problem by ranking teams based on a comparison of aggregated times. For example, Team A's aggregated time is always 83:20, while Team B's is 83:25 and Team C's is 84:55. Therefore, a ranking methodology based on the sum of times will always rank Team A ahead of Team B, Team B ahead of Team C, and Team A ahead of Team C.

Writing *Down* Your Thoughts

When I was a young writer for the Indiana University student newspaper (*The Indiana Daily Student* or *IDS*), a physics faculty member at the university refused to grant me an interview on the basis that "reporters simplify things to the point of fallacy." Though the man held considerable prejudice, he did hit on an important point. Citing the convention of the profession, our journalism professors instructed us to write at a level that the average ninth grader can understand. Such a restrictive rule eliminates all but the most elementary strands of logical essay. Further, it rules out any use of statistical analysis beyond mean, median, and mode.

Eventually, I left journalism to pursue economics. However, I have continued to think about journalistic conventions. As an avid news reader today, I sometimes come across blatant misuse of statistics in our media. Apparently, valid statistical interpretation comes after the ninth grade as well. I don't blame particular newswriters for these errors, however. After all, journalistic skills are a reflection of societal values. If we were a generally numerate society in terms of our demand of meaningful statistical analysis, journalists would provide the goods. However, the reality is that we are, for the most part, a numerically lethargic society willing to buy almost any statistical result that claims to be the result of rigorous evaluation.

In a *USA Today* article titled "Numbers can baffle in 5-setters," [6] sportswriter Douglas Robson writes, "When Roger Federer bounced back from a fifth set deficit vs. Rafael Nadal at Wimbledon in July, it represented a statistical aberration. It improved Federer's career five-set mark to a pedestrian 10-10." Firstly, Federer wins five-set matches fifty percent of the time. This is as much of a "statistical aberration" as a flipped coin landing heads up. Secondly, the record listed is not so bad, even for a top-ranked player such as Federer. In every five-set match, the two opponents have played each other evenly, in essence, over the first four sets. Therefore, it stands to reason that each player is equally likely to win the ensuing fifth set. In other words, Robson considers only a small set of Federer's matches in which he is, by definition, having a relatively hard time with his opponent. It should be no surprise that Federer's record over such matches is less than dominant. I can easily generate a statistic to make Federer look like an even less dominant player: Federer loses 100% of matches in which he loses the fifth set. The moral is that we should be careful that we are not selecting an outcome by our very choice of sample. When *sample selection bias* is present, it may not be the data in general, but the particular data that we select, driving the outcome.

Evidence of a Clutch Performer?

Upon Bob Knight's retirement from coaching, *ESPN Sportscenter* reported that the famed college basketball coach had accumulated an overall win-loss record of 902-371. They also reported that his win–loss record in contests decided by one point was 34-17. The anchor concluded of the latter statistic that all great coaches must be able to "steal" some close games. However, the statistic does not suggest any kleptomaniacal tendencies on the part of Knight. In light of his overall record, Knight's teams were much more likely to win, in general, than to lose. It follows that his teams would be favored to win by one point more frequently than to lose by one point. Knight's record in one point games is symptomatic of the overall quality of his teams rather than of Knight's supposed ability to pull out a close game. If anything, Knight's winning percentage in one point games (.667), when compared with his overall winning percentage (.709), indicates that he is perhaps *less* proficient in winning the close games.

But how can we assert that neither Knight's record in close games nor Federer's record in close matches is of statistical note? In the Federer story, the journalist chose to condition the statistic upon matches in which Federer was tied going into the fifth set. Following such conditioning, the selected matches are not likely to be representative of Federer matches overall. No such conditioning takes place in generating the Bob Knight statistic. We do not look at games in which Knight's team was tied entering the final minute of play. As such, we are broadly comparing the frequency with which Knight won by 1 point to the frequency with which he lost by 1 point. In general (i.e., over all games), we expect that a winning coach is more likely to win by some given amount than to lose by the same amount.

In these two cases, the samples have emerged differently. By interpreting the statistical output along with the sample selection methodology in each case, we find no evidence that Federer is bad or that Knight is good when the contest is close. Sports journalists and fans focus much attention on the concept of the clutch performance. Taking over a sports contest that hangs in the balance is widely viewed as an act of heroism rather than as a sign that the performer has a tendency to procrastinate, and much of sports fandom is based on the identification and subsequent idolatry of the modern-day hero. Sports fans often need to believe that an individual play is a reflection of character rather than a chance outcome. This is likely why the unverified clutch performer has such a prominent place in the sports pantheon.

Statistics and Consumer Behavior

An insurance company recently sent me an advertisement claiming that "seven out of ten customers who switched to [our company] paid less." The statistic was intended to make me believe that the company offers a lower price than its competitors most of the time. However, this statistic says as little about the company's relative pricing as the previous statistic said about Roger Federer's ability to win an average tennis match. In fact, the company's pricing claim could be true even if its coverage were cheaper to, say, only ten percent of all driver types. You see, a customer is more likely to switch insurance companies if the change would save her (i.e., a driver of her type) money. By virtue of the fact that the person has chosen to switch, it follows that she will likely save money from the action. This does not mean, however, that most drivers in general would save through such a switch. Who is doing the switching matters a great deal! This is especially true because firms of a given market often form niches in that market. For example, some insurance companies might specialize in insuring motorcyclists, while others might specialize in insuring truckers, and so on. Each such company might offer discounted insurance within their niche, as they possess certain efficiencies for providing insurance within that area.

This example relates to the concept of out-of-sample prediction in statistics. When we estimate a statistical relationship from a sample, we cannot expect that the estimated relationship will hold for additional observations that are outside of the range of the sample for a given characteristic. In the present case, the sample used by the insurance company can be summarized as *those who chose to switch to this company's auto insurance policy*. As I may not share any relevant characteristics with those who have previously switched to this company, the statistic related to saving money does not necessarily pertain to me. What saves money for the goose does not necessarily save money for the gander.

We can further illustrate the concept of out-of-sample prediction with a bit of absurdity. A team representative for the *Los Angeles Lakers* might find that LeBron James averages 31 points per game when he eats PB&J for lunch on game day and 24 points per game when he does not. Given this finding, an in-sample prediction might be that LeBron James will score 31 points the next time he eats PB&J for lunch on a game day. An out-of-sample prediction would be that Javale McGee will score 31 points the next time he eats PB&J for lunch on a game day. This prediction gives all of the credit for the 31 points to PB&J and none to LBJ. Without observing Javale McGee's performance level while on PB&J, moreover, we are predicting that he will attain the average in-sample performance. While this example is clearly absurd, it shows us how presumptuous out-of-sample prediction can be. We cannot use a sample from one population to say something about another population.

Motels and other businesses are also guilty of misleading consumers through the clever use of statistics. *Motel 6* often advertises rooms "from $59.99." In this case, the motel is characterizing its pricing by advertising only the minimum in a distribution of room prices. This minimum price may not provide an accurate representation of the motel's typical room price. Moreover, this price may not appropriately represent the price of a room that is typically *available* at the motel. Clothing stores advertise sales of "up to 80% off." In such a case, the store is characterizing a sale according to its maximum discount. However, the maximum discount might apply to only one item that was inexpensive in the first place. Imagine a store that sells 100 items. Ninety-nine of the items are regularly priced at $100 per unit, while the remaining item has a regular unit price of $1. Imagine that the $1 item is discounted to $0.50. The other ninety-nine items remain at their original price. Strictly speaking, this store can advertise that it is having a sale of "up to 50%," but the potential discount from this sale is only $0.50. Alternatively, imagine that the other ninety-nine items are marked down to $99.50, while the $1 item remains the same price. In this alternative case, the store can advertise savings of "up to 0.5%." This doesn't have quite the same ring to it. However, the potential discount from this sale is a full $49.50 ($0.50 per item multiplied by 99 items). Hence, the maximum discounted percentage is not always a good indication of a sale's overall magnitude.

By framing prices in a particular manner, firms are likely exploiting certain biases held by the typical consumer. That is to say, the store appears to be magnifying what the consumer thinks to be a selling point so as to get her through the door. Statistical analysis is one part mathematics and one part logic. It requires us to ask what type of data is important in a given setting and what this important data is telling us. Statistics never lie outright. They tell the knowledgeable user exactly what she sets out to find. However, those who misinterpret statistics may be at the mercy of anyone trying to sell them something.

References

1. Ellenberg, J. (2008). Vote! Why your ballot isn't as meaningless as you think. Retrieved from https://slate.com/news-and-politics/2008/10/why-your-ballot-isn-t-meaningless.html
2. CBS Sports. (2020). College basketball rankings: RPI. Retrieved from https://www.cbssports.com/college-basketball/rankings/rpi/
3. IAHSAA. (2019) Iowa high school athletic association 2019 football RPI rankings. Retrieved from https://www.iahsaa.org/wp-content/uploads/2019/10/RPI-Week-9-10.26.19-4A.pdf
4. Scott, J. (2007). The ratings percentage index myth. Retrieved from http://www.bigbluehistory.net/bb/rpi.html
5. Hammond, T. H. (2007). Rank injustice?: How the scoring method for cross-country running competitions violates major social choice principles. *Public Choice, 133*(3–4), 359–375.
6. Robson, D. (2008). Numbers can baffle in 5-setters. USA Today. Retrieved from https://www.pressreader.com/usa/usa-today-international-edition/20080121/282175056796172

The Things We Value

5

This chapter examines our valuations both in market settings and within society at large.

Why Don't You Two Get a Room?

In The Armchair Economist [1], an engaging book about the power of economic thinking, author Steven Landsburg asks the following question:

> Why is the second guest in a British hotel room charged a price roughly equal to that of the first guest, while the second guest in an American hotel room is charged considerably less than the first guest?

Landsburg leaves the question unanswered, an open economic riddle for his readership to tackle. A thinker of the noneconomic persuasion might conclude, "Different cultures just do things differently." Such arbitrary explanations of variation in everyday life do not please the economist, however. There is no value added in saying this. It is simply an ad hoc manner of explaining things that fails to identify the root cause. Moreover, one gains no predictive power through this assessment. Economists like to believe that people are largely the same but for the particular incentives that they face. Given these initial conditions, economists then like to figure out how different incentives might create behavioral differences in two settings.

Being an economist, I'd like to take a crack at *Landsburg's Hotel Puzzle* in the following pages. In addressing this puzzle, let us assume that people are amoral in their hotel dealings. They merely wish to gain the most consumer surplus (i.e., value derived from a good or service over and above its price) possible during their stay at a hotel, regardless of ethical considerations. If people are basically like this, and I know many who are, the hotel faces a problem in that a party of two may have an incentive to report only one guest so as to save on the cost of their room. Imagine that two buddies arrive at a hotel. (We must imagine because the following story is in no way autobiographical.) Buddy 1 steps into the lobby to check in, while Buddy 2 waits quietly in the parking lot. As part of the check-in process, the hotel attendant asks Buddy 1, "How many guests are with you, sir?" With the conviction of a B-movie actor, Buddy 1 replies, "It's just me." *Et voila.* Two people become hotel guests for the price of one. There are, of course, costs to underreporting the number of guests in a party. First, as long as the second guest remains undetected, the two buddies cannot stroll thoughtlessly around the hotel together. If detected, moreover, the pair will be fined an amount at least as great as the value of their underpayment.

As hotels are more efficient in providing hotel services than in detecting stowaways, they will tend to price the second guest such that the representative guest pair chooses to be honest. A guest pair will honestly report both parties whenever the expected cost of stowing away the second guest exceeds the hotel's price for a second guest. Hence, a hotel will set the second guest price marginally below the representative pair's expected cost of stowing that guest. As it becomes more costly to stow a second guest, a hotel will be able to charge a higher second guest price without encouraging additional stowaways. But what affects the cost of harboring a surreptitious second guest? One major factor is the size and layout of the hotel. For instance, it is much more costly to maintain a stowaway at a small hotel with one entrance than at a large hotel with many guests and multiple entrances. Detection is a great deal more likely at the small hotel, and this may be the source of the difference between second guest pricing in the United States and in Britain. On market average, British hotels are smaller and feature fewer entrances than their American counterparts. That is to say, they normally do not offer large, wraparound parking lots with a healthy choice of back entrances. Thus, it is relatively costly to maintain a stowaway at the typical British hotel. If both US hotels and British hotels charge a second guest price that is marginally below the expected cost of stowing away that guest, the British price is therefore expected to be higher.

Landsburg's hotel pricing problem is reminiscent of Becker and Stigler's malfeasant cop paper [2]. A malfeasant cop accepts bribe payments from those she catches violating the law. In exchange, she merely pretends that nothing happened. However, the cop takes a risk with each act of malfeasance in that she will be dismissed from the police force if detected. If a *bad cop* is detected by a *good cop*, for example, the *bad cop* will be reported. Becker and Stigler show that malfeasant policing can be eliminated by paying each cop the present value of her earnings in another profession (in the event of dismissal) plus the present expected value of her earnings from police malfeasance. In this salary structure, each cop is compensated as if she has accepted bribes. Consequently, no cop has an incentive to accept bribes.

In a paper written with Yang-Ming Chang [3], I studied the malfeasance of college basketball players. We concluded that "point-shaving" corruption by players (i.e., taking a bribe to influence a game's margin of victory) is encouraged by sub-optimal wages to players. In the case of college basketball, it simply does not pay enough for all players to be honest in light of the wages of corruption. The central moral of these stories is that people will behave legally, and sometimes even with something like social efficiency, if doing so is made to be worth their while.

This moral has implications not only for basketball game outcomes but also for issues related to law enforcement and economic development.

Oil Reserves

There is a long-standing political debate as to whether federally protected oil reserves, such as the *Alaskan National Wildlife Refuge*, should be opened to private drilling interests. Proponents believe that such drilling ventures will increase supplies of oil and push down world oil prices. Many detractors suggest that it will take years in drilling and pipeline development for such new oil supplies to become available to the market. Therefore, detractors argue, the present price of oil will be unaffected by such a policy change.

There are many issues to consider when deciding whether to open protected lands for commerce. One important consideration is the amenity value of such lands, which often possess rare and beautiful habitats. This value, which persists over time if unspoiled, can be compared directly to the economic benefits of drilling, where such a comparison aids in making a policy decision that is in the best interest of society. In an academic article [4], economists Morris Coats, Gary Pecquet, and I apply a result first developed by pathbreaking economist Harold Hotelling in 1931 [5]. Doing so, we find that the present and future prices of oil are fundamentally related given that oil is nonrenewable. Specifically, an increase in the future supply of oil pushes future price (profit) expectations for oil companies downward. In the presence of lower future expected profit, the opportunity cost to firms of supplying market-ready oil in the present (i.e., not holding such oil for future sale) declines. Thus, firms supply more oil in the present, and the present price of oil decreases. To put the scenario more succinctly, an opening of oil reserves is expected to decrease the present price of the good, even if the reserves in question would not directly supply the market until years hence.

Why has this important and long-standing result not entered public discourse on the opening of protected oil reserves? Perhaps people are accustomed to thinking about goods and services for which production is non-rival across time. A good is non-rival across time if current production and consumption of the good does not crowd out future production and consumption. As distinct from the case of a depletable (but storable) resource, the current production of a non-rival good does not constrain future production. Therefore, there is no intertemporal production scarcity to discourage the present production of such goods and services.

As an example, we can consider a government that announces its intention to lift a binding ban on income tax preparation by foreign firms. Assume that the new policy will be effective 1 year from the date of the announcement. Such a move shifts the expected future supply curve for tax preparation within the country, as foreign tax preparation firms will contribute to future supply. Thus, the future price of professional tax preparation is expected to decline in the country. However, people cannot simply hold off on this year's tax return and prepare twice as many returns next year. As production of tax returns is non-rival and non-storable, present domestic tax preparation decisions are not influenced by the future policy change. For the time being, domestic tax preparation firms and their clients will continue about their business as if the announcement was never made, and the present market price of professional tax preparation will not change in response to the announcement.

Along these same lines, we might imagine a small, isolated town that possesses one restaurant. The town board decides to rezone a five-acre plot of land, effective 1 year hence, such that it can be used for commercial purposes. At least two more restaurants are expected to open once the rezoning becomes effective. Thus, we expect a higher future supply of restaurant services in the town at a lower future price. Current restaurant service production in the town does not rival future restaurant service production. The townspeople cannot simply purge until next year and then binge at the lower market prices. As such, the present prices and quantities of restaurant services remain the same.

As one final example, the late economist Morris Coats pointed out that an increase in the number of accredited medical schools in the United States would increase the supply of doctors 10 years hence but would not be expected to drive down the present price of medical treatment. There aren't many ailments for which a person could wait 10 years to treat. If you actually broke a leg during your first acting audition, for example, you probably wouldn't just sit in a chair for the next 10 years and wait for a flood of new medical doctors to enter the labor market. Rather, you would likely rush off to a medical facility before gaining any information about future changes in the supply of doctors. There are many additional examples of markets in which expected future price changes do not affect current production decisions. When discussing intertemporal price relationships, we must treat oil and other depletable resources as a special case of goods.

Ode to Slightly Less Able People

There are people who are disabled and then there are those who are slightly less able. *Seinfeld* character George Costanza was a *bona fide* member of the latter group. He was slightly shorter, slightly less intelligent, slightly less contented, and had slightly less hair on his head than the typical man. In an episode of the show, George himself says, "I'm disturbed. I'm depressed. I'm inadequate. I've got it all." George felt that the sum of his shortcomings entitled him to disability status. This belief led him to shamelessly pose as a disabled man to obtain a good job. In his slightly ridiculous way, George had a bit of a point. We have all sorts of laws protecting disabled individuals. Public buildings must feature disability accessible entrances, and disabled individuals receive "equal opportunity" in job hiring decisions. Organizations representing disabled people have lobbied effectively at all levels of government such that the opportunity set for this group more closely matches that of the able majority. George's group of slightly less able people, though slightly disadvantaged in so many ways, is generally neglected under our legal and cultural systems.

As an example, I once boarded a flight on *Southwest Airlines*. After taking my seat, I noticed that another passenger, a petite female, was struggling to hoist her heavy but allowably sized carry-on bag into an overhead compartment. She snatched and lifted the bag once but was able only to get it to the level of her chest. She tried again and barely raised the bag above her shoulders. Another feign attempt sent the bag to the level of her waist. I sat nearby, bemused by the young lady's struggles, and rejoiced in my adequate level of strength. (Actually, I wanted to help her but did not have access to the aisle.) Eventually, a flight attendant came by, snapped up the bag, and stowed it for her. Afterward, the attendant, obviously miffed by the ten seconds of customer service she had provided, said to the female passenger, "You should never bring a carry-on that you are unable to stow by yourself!" The statement rung in my ears in the way that inconsistent statements sometimes do. A company that ostensibly prides itself on service to disabled passengers was openly discriminating against a slightly less able, though not legally disabled, person. By virtue of genetic factors, the flight attendant was contending that this passenger was entitled to slightly less baggage weight than other passengers.

This episode stuck with me and caused me to wonder why slightly less able people (e.g., the thirtieth percentile person in strength or intelligence) are commonly held to the same standards as everyone else, while disabled people are protected in so many ways. I developed two plausible explanations. First, we can effectively implement a special status law only if we have the ability to identify qualifying individuals. If ability level, in various respects, follows an approximately bell-shaped distribution, as is the case with intelligence, height, and many other physical characteristics, then a large proportion of individuals are close to average. The mode, or most likely value, in a bell-shaped distribution is at the distributional average. Therefore, the bulk of the population is far removed from any sort of disability threshold but relatively close to slightly less able person status. For most individuals, therefore, the cost of credibly faking less able status is lower than the cost of faking a disability. As such, we would expect a greater number of less able person imposters than disabled person imposters for a given law. In this sense, laws protecting less able people are unlikely to be as attractive to policymakers in terms of the ability of such laws to create social value.

Further, slightly less able individuals have weaker incentives to organize and lobby for special protections. Consider two individuals, a disabled person and a slightly less able person, both of whom buy one plane ticket per year. The disabled person is not able to enter a plane unassisted or stow her bag properly. The slightly less able person has no trouble entering the plane but is unable to stow her bag properly. Assume that each person values the assurance of plane entrance at $50 and the assurance of bag stowage at $25. As such, the disabled person values legal protection at $75, whereas the slightly less able person values legal protection at only $25. The disabled individual will spend up to $75 (in time, effort, and cash payments) lobbying for legal protection on planes, whereas the slightly less able person will spend up to $25 in the same pursuit. Thus, disabled people have stronger incentives to lobby for special protections. Lastly, it will be easier for disabled people to recruit advocates in the fight for certain legal protections. Their need for

such protections is often unequivocal. That they enjoy these protections to a greater
extent than the slightly less able is no surprise.

The Computers Are Watching You

From *Burger King* to the *Hard Rock Café*, computers have saturated the restaurant
production process. The question remains, "Why are they so important to the
restaurant industry?" Their basic purpose is to transfer orders from the waiter, or
cashier as it were, to the cook. However, pencils and notepads do the same job
almost as quickly. Further, pencils and notepads are less expensive and more
reliable, in terms of downtime, than are computers. So why was the restaurant
industry in such a big hurry during the 1990s and 2000s to update the old
handwritten system? Economist Alan Blinder finds that the adoption of information
technology by businesses in the 1970s and 1980s did not immediately help worker
productivity. He attributes this outcome to the misapplication of information technology by businesses during that era. Many businesses of that period appear to have
put the cart before the horse when integrating computers. Blinder states that, while
many businesses became excited about the ability of computers to improve production processes, fewer understood how the new technology should be applied.

Could it be that much of the restaurant industry remains in ignorance even today,
using computer ordering systems when paper and pencil ordering systems work at
least as well? This does not seem to be the answer, as industries have long since
approximated the proper role and scope of computer technology. Computer ordering
systems are valuable in the restaurant industry not so much because of their
productive efficiency compared to handwritten orders. They are valuable because
they hold cashiers and waiters accountable. Using pencil and notepad, a cashier can
easily embezzle money, for example. She can do so by writing down a full order on
the notepad, handing that order to the cook, charging the customer full price for the
meal, and failing to ring up all or part of the order on the cash register. Using the old
system, a cashier need not provide the same information to the cook as to the cash
register. As such, she possesses billing discretion under the old system. Though the
restaurant might eventually find that inventories aren't matching up with receipts, it
is difficult and costly to detect why this is so. Thus, the pencil and notepad system is
not very useful toward monitoring conduct.

Computers, on the other hand, integrate the relaying of the order with the
charging of the bill. They can even identify which employee is placing the order.
In order for a chef to know what items to make, the cashier must generate and send
an order to the chef. That order is linked to a bill for the customer that must bear a
charge for each ordered item. In essence, the restaurant computer system's value
appears to be derived from its ability to minimize embezzlement opportunities by
removing billing discretion from restaurant employees. Bucking the industry trend,
many mom-and-pop restaurants, in which employees are members of the same
family, have not moved to any sort of computerized system to relay orders. This is
likely because such firms carry a much lower embezzlement risk than their less

family-centered competitors. In many such restaurants, each employee has something akin to ownership interest in the restaurant, as restaurant profits directly increase family income.

The present essay was inspired by economist Robert H. Frank. In a column for the *Huffington Post*, Frank describes an interesting student essay that explains mandatory *receipt* policies at restaurants. Such policies, the student argues, are meant to hold the cashier accountable. If a cashier is bound to return a receipt to each customer, then her opportunities for embezzlement—by misrepresenting the transaction at the register—are greatly diminished. In adopting such a policy, a restaurant is essentially enlisting the involvement of customers in monitoring employees. It is likely that computer ordering systems and mandatory receipt policies both seek to address the same issue of restaurant employee accountability.

Other Forms of Workplace Behavior That Can Be Explained with Economic Analysis...

While making sense of workplace behavior, we can address the age-old issue of doctor handwriting. The general illegibility of doctor handwriting raises several questions. *Is the medical profession, in general, a population of highly intelligent people who don't know how to write? Does a doctor feel more oracular if patients cannot possibly decipher what medicine she is prescribing? Are pharmacists required to take a handwriting decryption class that is unavailable to the general population?*

Compare the scribblings on a typical medical prescription with the latest handwritten letter from your grandmother. It is not just that *your* grandmother probably has excellent penmanship; I'm willing to bet that more than 99% of literate grandmothers do. But if this is the case and essentially all grandmothers can achieve scribe-like results with the pen, why are doctors having such a difficult time stringing together six legible words? It is probably not the case that penmanship is beyond the ability of most doctors. Rather, it is likely that most doctors write so poorly because they have a high opportunity cost of time. The typical doctor is worth several hundred dollars an hour in terms of the value of services that she provides. If a doctor can save even an hour per week by writing in a *barely* legible manner, this is a great boon to society. Such a doctor is suddenly able to treat a few more patients each week. Your grandmother, on the other hand, has a low opportunity cost of time. It isn't nearly as costly for her to employ good penmanship in writing a letter to you. The extra time she spends on penmanship might deprive her of one episode of *Wheel of Fortune* per week, but watching the show is probably more of a habit than anything at this point. We can assume that she stopped thinking of Pat Sajack as a dish several years ago and that she has since developed a sinking feeling that the wheel is rigged for the purpose of suspense.

Economists believe that a person's behavior is largely determined by the incentives she faces. There is no cohesive accounting for other potential behavioral drivers. Whereas doctors face a high opportunity cost of penmanship, your grandmother's opportunity cost is relatively low. If your grandmother is a former doctor, I'm willing to bet that her penmanship made a miraculous recovery upon retirement.

Prices Make Firms Do Funny Things

During the Spring of 2008, the world price of rice increased dramatically. The staple's price tripled between November 2007 and April 2008 according to a USDA report [6]. In response, stores such as *Costco* and *Sam's Club* implemented rationing policies, whereby individual purchases were not allowed to exceed a specified quantity [7]. At the time, there were signs in *Costco* stores that read, "Due to the limited availability of rice, we are limiting rice purchases based on your prior purchasing history." The stores claimed to implement such a policy so that rice would be available to all customers. However, it seems strange for a firm in a competitive industry, in any industry for that matter, to pass up potential sales in favor of social equality. Stores that face increases in the price of merchandise typically transfer price increase to customers. Thereby, the price mechanism acts as the rationing mechanism rather than any other store policy. The curious policies of *Costco* and *Sam's Club* in this case caused me to consider if there might be a more fundamental reason that such store policies were enacted.

One plausible explanation is that *Sam's Club* and *Costco* wanted to hold current inventories of rice while the good's price continued to rise. As the old business mantra goes, "Buy low and sell high." This mantra implies that one should hold if the price is getting higher. These two wholesale stores are particularly suited to limit current sales through purchase history rationing in that they necessarily hold the purchase history of every customer. Therefore, it is quite costly for a customer of *Costco* or *Sam's Club* to obtain more than her ration. At these stores, the customer who wishes to evade rationing must find another customer willing to purchase additional rice for her. If rice market speculation served as the underlying motivation for these store policies, why did the stores not simply fail to restock their shelves with rice? After all, such an alternative store policy would have been at least as effective in allowing the stores to hoard rice. However, it would have reflected poorly upon their organizational ability and might have done more harm, in terms of customer relationships, than good to the bottom line. By rationing, *Sam's Club* and *Costco* were able to project an egalitarian view of customers and maintain their perceived level of organization, all while making a potentially hefty profit on future rice sales.

Modern Gift Giving

Gift giving can be a risky endeavor. It is often unclear to the giver what the recipient already possesses in the way of material goods, what she wants, and what others will buy her. Gift registries overcome these basic problems by providing information on the recipient's wants and listing which of those wants have already been fulfilled by others. There is one problem, however, that the conventional registry does not fully overcome. Suppose, for example, that a recipient values a particular $200 item by much more than any possible combination of four items priced at $50. In other words, the recipient would rather receive four quarter-shares of the $200 item than any four products worth $50. However, if the recipient believes that her benefactors are unlikely to coordinate on a large purchase, she may not list the $200 item at all for fear of appearing avaricious.

And therein lies the frequent failing of the typical gift registry. It cannot be used to coordinate group purchases and thus limits the polite recipient from listing higher-end items. Although gift-givers are able to use a registry to communicate with one another toward the coordination of a group purchase, it is often difficult for individuals inclined toward a particular product to find one another. In the case of a wedding gift registry, members of the groom's party may not be familiar with members of the bride's party and *vice versa*. An uncoordinated group purchase mechanism, however, would allow respective gift-givers to purchase and receive credit for purchasing shares of items at different points in time and without direct communication. In doing so, such a mechanism would allow for many more group purchases by eliminating the associated search and coordination costs. A share purchase option is essentially a way of efficiently matching complementary gifters.

Such a mechanism would work as follows. Let Fred be the first gifter to visit a particular registry. He sees that the registry contains two items, a $50 item and a $100 item that can be purchased in shares. Fred decides to buy half of the $100 item. Jennifer is next to visit the registry. She sees that the registry contains two items, a $50 item and half of a $100 item (as Fred has bought the other half). She decides to buy the remaining half of the $100 item. The $100 item is now fully purchased and no longer available in the registry. Lisa is third to visit the registry. She sees that the registry contains only the $50 item and decides to buy that item. Obviously, there is the risk that an item is never fully purchased. In such a case, the registry could reallocate unused funds according to the recipient's wishes. The registry could even be set up such that the recipient could buy out the unpaid share of a larger item. If, for example, gifters have paid for $175 of a $250 item, a "buy out option" would allow the recipient to pay the remaining balance of $75 and receive that item.

As it stands, the recipient can always exchange gifts according to her tastes. However, she may not wish to do so if the gift-giver would notice that the original gift is absent. The share purchase option would increase benefits accrued by the recipient for a given number of dollars spent. It would do so by encouraging her to ask for those items she most wants, regardless of their price.

A Shutdown Rule for Marriage

Divorce is stigmatized in most cultures. Many people believe it to be a form of failure. Economists generally view divorce as a triumph of marginal decision-making. That is, it is a natural response to the realization, by at least one party, that the additional benefits from remaining married are outweighed, in expectation, by the additional costs. According to *The Economist*, the United States has one of the highest divorce rates in the world [8]. Does this suggest that American values are somehow lacking? It certainly does not. Rather, our inflated divorce rate indicates that marriage laws here allow for a freer, and therefore more accessible (i.e., less difficult to obtain), market for divorce. Greater accessibility causes an increase in the quantity demanded of divorce, as per the *Law of Demand* in economics. Beginning in 1969 with the passing of the *Family Law Act* in California, legislation allowing for no-fault divorce has become increasingly prevalent in the United States. By 1985, every state in the United States permitted no-fault divorce. However, such laws have yet to scratch the legislative surface in many developing countries, where divorce is said to be less of a "problem."[1]

In a free society, one can reason that divorce is a good thing, on net, whenever it happens and a bad thing, on net, whenever it does not take place. Mutually agreed upon divorces are *Pareto Improving* (beneficial to both parties) in the sense that each party has revealed a preference for being divorced. Thus, the joint well-being of the two individuals is higher after separation. However, what if only one person wishes to become divorced? In this case, joint well-being is maximized under divorce only if the alienated spouse values divorce by more than his or her partner values the marriage. Does this mean that some one-sided divorces decrease the couple's total well-being? The answer remains as *no*. Even a one-sided divorce must be efficient in order to take place. This is true because there is an internal bargaining mechanism in each marriage. Gifts, money, quality time, family planning, and freedom are freely exchanged as marital bargaining chips. If an alienated partner values divorce by less than his or her spouse values marriage, the latter can offer the former a "bribe" sufficient to keep him or her in the marriage.

A token marital bribe in the United States, for example, is that the husband promises to do the dishes more often. Whether the husband should have been doing the dishes regularly in the first place is beyond the scope of our present analysis. Provided that the bribe is accepted, both parties are better off than if a divorce had taken place. They have revealed a preference to stay in the marriage under the new terms of the agreement. And you thought marriage was romantic. Whenever a one-sided divorce does occur, it is because the alienated partner values divorce by more than his or her spouse values the marriage. In this case, there is no possibility for mutually beneficial bargaining, and the alienated partner optimally files for divorce. In other words, we can be sure that even a one-sided divorce, when it occurs, is in the best aggregate interest of the involved parties.

But why does divorce happen in the first place? In a free society, after all, members of a divorced couple once thought enough of each other to file for a joint mortgage. They coined ridiculous nicknames for each other and coerced acquaintances to devote a day toward observing their union. So what happened? Much like a used car, marriage is an experience good. In the case of such a good, it is

---

[1]What is the unhappy marriage rate in such countries?

difficult for the consumer to assess quality until after purchase. He or she must *experience* the good to make such a judgment and thus faces uncertainty at the time of transaction. The consumer's marginal decision might subsequently change due to adverse experiences.

A potential spouse is an experience good in the sense that one doesn't know what kind of marriage partner the spouse will be. You might have a very good idea about the quality of your boyfriend *as a boyfriend*. However, marriage presents a challenging set of circumstances, and dating can only signal how a prospective spouse might behave in the artificial arena of dating life. Marriage consumers negotiate this uncertainty by attempting a marriage and observing the nature of the spouse over time so as to collect additional information. In light of how marriage is used in the real world, then, it might be more descriptive for wedding vows to read, "...until optimal decision-making do us part."

The Economist as Interior Designer

In my former office building, there is a copy room. The room features a large picture window that overlooks a pleasant campus courtyard. It took me about 4 weeks to notice this view because I, as most people, usually entered the copy room in a hurried state. In the same former office building, there is also a faculty lounge. The faculty lounge is in the center of the building and thus offers nothing in the way of a view. It took me about one minute to notice this absence of view because I, as most people, usually entered the lounge in a relaxed state. One day, while drinking a soda and staring at the blank lounge wall, I realized that an economist could not have assigned rooms within the office building. An economist, trained in comparing the efficiency of different decisions, would have recognized the extreme scarcity and value of common window space in the building. Consequently, she would have allocated this window space so as to maximize the collective benefit of the building's faculty and staff.

A picture window is more beneficial when a person is on break than when the same person is in a hurry. In the former state, the person is generally more inviting toward and cognizant of aesthetic distractions. Thus, an economist would have designated the picture window room as a faculty lounge and the center room as a copy area. I tell this story to make the point that economics is just as applicable to subjects such as interior design as it is to monetary policy. Economics studies decision-making under situations of scarcity, and people make such decisions in a variety of settings.

Accounting for Life and Death

Disasters are, by definition, horrible events. Yet they are seldom as bad as reported. The economic cost of a natural disaster, such as a hurricane landfall, can be thought of as the event's explicit cost minus the explicit cost that would have been incurred had the event not taken place. When a newspaper article states that an ice storm has cost a region ten driving fatalities, there is no accounting for how many road fatalities would have occurred in the absence of poor road conditions. In other words, ten lives is the explicit cost of the ice storm rather than the economic cost. Let us assume there are ten road fatalities over some area during a weeklong ice storm, but two road fatalities were expected to occur over the same area and time period in normal weather conditions. In this case, the ice storm has taken $(10 - 2 =)$ 8 lives rather than 10.

A similar reporting error is often made in estimating civilian death tolls during wartime. Some such counts used by media sources fail to consider the simple fact that death would occur even in the absence of war. If a media outlet wishes to depict a certain party to the war as heinous, the explicit human death toll becomes a popular figure. *Iraq Body Count*, an independent team of academics and activists in the United States and United Kingdom, spent years diligently reporting the number of civilian deaths from violence related to the Iraq War. Though the team painstakingly counted documented civilian deaths from war-related violence, their methodology tells us only the explicit human toll of the Iraq War, unadjusted for the effect violence would have had upon Iraqi civilians without war. Such a count cannot truly assess the degree to which Iraqi citizens are affected by the war. Economist Joel Potter points out that heat wave death tolls among the elderly are often overstated in that some positive fraction of those elderly people were expected to die even if a heat wave had not struck. When gauging the toll of a natural disaster, we must compare what happened with what was expected to happen. Often, we instead compare what happened to what would have happened in a perfect world. This line of reasoning may seem insensitive. On the contrary, however, we stand to save more lives through policy if we account for the true risks of each given event. Only by doing this will we understand those policy choices that will truly save the most lives.

On the lighter side of economic costing, my treadmill displays the number of calories that I've burned through exercise. This figure, which is based on an estimate of the energy that I've generated, is not a very good indicator of how much better off I am through exercise. The fact that I burn calories even while sitting on my couch indicates that some of those calories would have been burned simply as a consequence of my existence. In this sense, the machine's estimate of calories burned overstates the number of calories lost from exercise. Conversely, aerobic activity thrusts me into a state of high metabolism even in the minutes of inactivity following exercise. In this sense, exercise causes me to burn *more* calories than the machine estimates. Given these countervailing omissions, it is hard to tell exactly how many calories an individual actually burns as a consequence of exercise vis-à-vis the machine's estimate. It could be considerably less or more than advertised.

The Opportunity Cost of a Trade

The opportunity cost of a decision is the foregone value of the best alternative to that decision. If you value the experience of watching *Monday Night Football* at $50 and the experience of spending Monday night with your wife at $40, then your opportunity cost of watching *Monday Night Football* is $40, and your net economic gain from the activity is $50 – $40 = $10. Sportswriters often neglect the concept of opportunity cost, especially in analyzing player trades. For example, *ESPN* sportswriter J.A. Adande wrote in 2007, "The *Lakers* have to trade him (Kobe Bryant). They can't let one of the league's top players walk away (in the summer of 2009) and get nothing in return." Adande believed that the *Lakers* should have traded Kobe Bryant rather than keep him and allow him to opt out of his contract 2 years hence. The *Lakers* should have done this, according to Adande, to avoid receiving no compensation for his exit. However, the *Lakers*, who obviously did not trade Bryant at the time, *were* compensated in that *Bryant remained on the team for the remaining 2 years of his contract*. That is, the opportunity cost of trading Kobe Bryant in 2007 would have been not having Kobe Bryant on the roster for the pursuant two seasons.

If the team had traded Bryant at the time, we expect the present value of the player(s) the *Lakers* would have received in the trade to roughly equal the present value of having Kobe Bryant on the *Lakers*' roster for two more seasons. Otherwise, why would several well-informed *NBA* executives agree to the trade? Therefore, it is not clear that the *Lakers* would have been better off in trading Kobe Bryant. In the case of sports trades, it is not players that are being traded but the remaining terms of player contracts. Economist Joel Potter points out that another common fallacy when discussing sports player trades involves the consideration of sunk costs. Sunk costs are expenditures that have been incurred and cannot be recovered. Unlike an opportunity cost, a sunk cost is irrelevant to present decisions. An oft-repeated lament among baseball fans goes something like this: "We shouldn't trade away our young prospects because they've taken years for us to develop." However, the amount of resources it required to develop a player is irrelevant in the present, as development costs cannot be recovered once incurred. As mentioned in the previous paragraph, there are two important considerations when a team decides whether to trade a young prospect. The first consideration is the team's present value for that prospect under the terms of her present contract. The second consideration is the team's present value for the player on the other end of the trade under the terms of her present contract.

Assume a major league baseball team has a young outfielder who has gone through its minor league system. They presently value this player's future services at $12.5 million. The resources necessary to develop the player to this point cost the team $600,000. Suppose the team has an opportunity to trade the first player for another young player who has an identical contract in terms of salary and number of years. While not necessarily more talented, this second player better fits into the team's lineup. Therefore, the team values the second player at $13 million. Should the team make a trade or stay with the same player? If the team recognizes its sunk costs and erroneously values the first player at $13.1 million ($12.5 million in present value plus \$0.6 million in sunk development costs), it will choose to stay with the first player. However, the team will lose in the sense that it receives only \$12.5 million of expected value rather than \$13 million. The team should make this trade. The \$0.6 million in development costs have been incurred. They have already affected, in expectation, the present value of the first player. Therefore, considering such costs in any present decision framework represents a double counting of the player's value attributable to developmental investment. Decisions regarding a player's future area code should be based simply upon the amount by which a team values the player's services vis-à-vis the services of other players *going forward*. When acting the general manager, baseball fans should remember to forget costs that have sunk below the decision surface.

Gambling with Tax Dollars

The United States is amidst a higher education bonanza. College participation rates among citizens aged 18–24 years have risen substantially over the past several decades according to the *American Council on Education*. Blanket government subsidies have greatly expanded the function of our higher education system. Once a haven for the young elite, our contemporary higher education system serves a much broader societal function. There is a strong economic argument in favor of such subsidies. If higher education creates external benefits in the labor market, a government wishing to improve social well-being should support its pursuit to some degree. That is to say, if my higher education not only causes me to be a better worker but also my (educated or uneducated) colleagues, then defraying the cost of student tuition can benefit society.

However, numerous economists and policymakers believe that these subsidies have become indiscriminate in identifying potential post-secondary students who actually want to participate in the academic portion of college. The belief that higher education has *external* benefits in all cases presumes that each student gains human capital from the experience. As a university professor, I've observed many cases in which a freshman rarely attends class and subsequently fails out of school in the first year. Society loses in each case of such subsidies in that these students ostensibly gain little or nothing academically from the experience. Therefore, there is little or no external benefit from subsidizing an individual whose high school performance suggests that she is ill-equipped or ill-motivated for further study. Moreover, the ill-equipped, ill-motivated student could use the time following high school graduation to either develop skills in a suited field or to gain experiences that may mold educational values. J.D. Vance, acclaimed author of Hillbilly Elegy and *Yale Law School* graduate, attributes his time in the military as helping to refine his educational perspective.

Let us consider an example in which there are three people contemplating college enrollment. Each person costs the same amount, say \$3000, to subsidize under the government's higher education plan. The subsidy reduces the private portion of each student's yearly tuition fee from \$10,000 to \$7000. Student A stands to gain a great deal from college studies such that her present value of higher learning, in terms of increased real lifetime income, is $11,000 per year. Given external benefits from education, the total social benefit from Student A's higher learning (i.e., the benefit to her *and* her future co-workers) is $12,500. Student B stands to acquire less human capital from education such that she values higher education at $9500 per year, and society values her attending college at $10,500 per year. Student C is not suited to gain as much from college as her peers. She values post-secondary education at $7500 per year, and society values her attending college at $8000 per year.

We can now sort out the winners and losers in each of the three subsidy cases. The first subsidy is merely a transfer from taxpayers to Student A. Student A gains $3000, and taxpayers lose the same amount. The student would have pursued a degree even in the absence of a subsidy because she places more private value on a college education ($11,000) than even the *unsubsidized* tuition rate ($10,000). Therefore, society is roughly unchanged in the case of this subsidy. Student B is also helped by the subsidy. She would not have attended college at $10,000 per year but gainfully enrolls at a subsidized cost of $7000 per year. Student B is better off by $2500 ($9500 minus $7000) in attending college under subsidy, and society as a whole is better off by $500 ($10,500 in total benefit minus $10,000 in total cost) for having her attend. Student C would not have attended college at $10,000 per year but enrolls at a private, subsidized cost of $7000 per year. As her benefit from college education is $7500 per year, she is better off by $500 annually in attending college under subsidy. However, society as a whole is *worse off* by $2000 a year ($8000 minus $10,000) in helping Student C attend college.

It could be that most students accepted into public universities are of the A or B type. However, more research is necessary to ascertain the discretionary level at which tuition subsidies do the most social good. As in the case of many government-funded scholarships and grants, it appears that subsidized student loan access should be granted on a more meritorious basis such that it targets productive students who will be able to return the social cost of the subsidy in the form of external productivity benefits. Another important consideration is whether the government should subsidize students over every course of study. As discussed earlier, some college degrees build human capital and also sort students into ability groups, while others are useful primarily for their ability to sort students alone. If a course of study isn't particularly effective in building human capital, the benefits from subsidy may not justify the social cost.

Ignoring Your Team Like a True Fan

Loyal fans, those who attend games whether their team is having a winning or losing season, hurt the team's future success. The more of them, the worse team will do in the future. Fair-weather fans on the other hand, those who flee when the season begins to get rough, help to ensure a more successful future for their (sometimes) team. Consider a place (Wrigleyville, Chicago) in which fans support their baseball club (the *Chicago Cubs*) through thick and thin. The team carries on strong

attendance in good years as well as in bad. Consider another place (Miami), in which
fans abandon their team (the *Florida Marlins*) at the first sign of adversity, only to
shamelessly re-emerge when the club is winning again.

Now let us ask the question: Which team's owner has more incentive to build a
winning team? Is it the first, whose profit margin is tied loosely to the team's winning
percentage, or the second, who earns a profit largely by producing wins? *Marlins* fans,
by showing up when the team is up and disappearing when the team is down, are
providing stronger incentives to the owner of their team than are *Cubs* fans. This is not
to say that we necessarily expect the *Marlins* to be better than the *Cubs* in any given
year. There are other factors at play in determining win performance, such as market
size, managerial efficiency, and chance. However, it is to say that the fair-weather fans
of Miami are expected to motivate more additional expected wins for their team than
are the rock-solid fans of Wrigleyville. Perhaps your father used to tell you that
everyone needs a little tough love sometimes. The Marlins are apparently no exception, having parlayed erstwhile fan support into two *World Series* titles during their
brief franchise history. In a classic sports economics article [9], economist John
Vrooman shows empirically that teams whose profit margin is sensitive to winning
percentage tend to perform better, other things being equal, than those with loyal
attendance numbers, even when controlling for such factors as a team's market size. In
fact, this result holds for all major North American sports leagues.

To amend an old adage, it can be said that quitters never win, but teams whose
fans quit during losing periods probably win more in the long run than they
otherwise would. Before we criticize another fair-weather fan for jumping on the
bandwagon at just the right time, we should think of her long-run contribution to a
winning team culture.

Nomadic Traders

The halls of academia are occasionally visited by people of some practical importance. These people are on the perpetual hunt for textbooks but care not of the
knowledge therein. They, the textbook middlemen and middlewomen, buy books
from professors (who previously received them from publishers) and resell them to
bookstores. Many an honest textbook middleperson has informed me that certain
universities are openly hostile toward their kind. A dean or department chair might
instruct faculty members not to deal with textbook middlepeople. In other cases, an
administrator or faculty member might "ask" such a person to leave the premises
altogether. As a graduate student instructor, I was told by more than one faculty
member to avoid anyone who visits campus with an *ISBN* scanner. Of course, I
didn't heed this recommendation. It seemed hypocritical to me that certain economics professors, for example, could tout the benefits of market exchange in class but
openly seek to restrict a textbook market once the lecture was done. Further, it struck
me as short-sighted for university officials to discourage such market activity.

In essence, the textbook middleperson subsidizes faculty salaries for the university. If this sounds like an odd conclusion, consider the effect of the middlepeople

from a market perspective. They provide a faculty member with an income-enhancing opportunity that is directly linked to that professor's university position. That is, the professor would not possess the inventory or location to sell textbooks without being employed as a professor. And there are many "dissident professors" who take advantage of bookselling opportunities. In this sense, the middleperson increases the expected gains associated with being a professor. These increased gains, in turn, slightly increase the supply of professors, which drives downward the equilibrium market wage for a professor. While this effect may not be huge, it certainly calls into question why a university official would take issue with an unobtrusive textbook middleperson.

Sticking It to the Mother Country (Again)

Baseball and cricket have many similarities. They are both bat-and-ball games played on a grass field of varying dimensions. Further, both sports originated, more or less, in England. As far as we know, cricket was developed by English shepherds who used their crooks as bats, while baseball was developed in the United States from an English stick-ball game known as "rounders" (a name that alludes to the baserunner's "rounding" of the four bases). While cricket enjoyed some level of popularity in the 19th century northeastern United States, the US moved decisively toward a rounders-style game just as England began moving away from it. What a coincidence! Thus, today we find that most of the more recent British colonies play cricket, while countries currently or formerly under the hegemonic rule of the United States (e.g., Japan, South Korea, and Latin America) have tended toward baseball.

It is bizarre that there exist two international games that are so similar but so culturally separated from one another. With the exception of Canada and Australia, two countries that broke from the British Empire neither very early nor very late, no country has participated in both the *World Baseball Classic* and the *Cricket World Cup*. If a fan of one sport considers the other at all, it is usually to deride the other sport for its relative simplicity or dullness. I once endured a group of avid cricket fans who felt the need to confront me with evidence of baseball's obvious inferiority for several minutes. At length, their credibility as expert critics was lost when they asked me how exactly a run was scored in baseball.

When I began watching cricket matches several years ago, I recognized many elements of the game as analogous to those that comprise the American national pastime. However, I didn't completely understand the batting stance of the typical cricket player. Instead of employing an upright stance and a fully cocked bat, as is seen in baseball, the cricket player employs a slightly more anemic-looking position, crouching forward slightly and keeping the bat low and only slightly cocked. Those readers with a good knowledge of professional baseball might think of the opposite of Julio Franco's stance [10]. After watching cricket matches for some time, the stance began to make sense vis-à-vis the baseball stance to which I had grown accustomed. The cricket bowler typically skips the ball low off the ground with the intention of hitting any part of a wicket that extends just 9 in. above the ground.

Whereas baseball features a chest-to-knee strike zone, the cricket batter has what is essentially a shin-to-foot strike zone, where the ball is allowed to bounce in cricket. These differences force the cricket batter to hit on a different plane than the baseball batter. Whereas the baseball batter adopts a swing that is essentially parallel to the ground, the cricket batter swings in a way that more closely resembles the middle portion of a golf swing. As in the case of a golfer teeing off, a cricket batter's premium air space is more easily accessed from a slightly crouched position.

But why doesn't the cricket batter cock the bat more with the purpose of swinging more forcefully along the same plane? Is there no Julio Franco plate appearance footage available in cricket-playing countries? This is where economic thinking comes in handy. The cricket batter is again responding to costs in choosing not to cock the bat. If one passed ball hits the wicket, the cricket batter's turn expires. On the other hand, if a baseball batter allows a strike to cross the plate, he or she receives additional opportunities to put the ball into play. In other words, the cricket batter must adopt a more defensive stance precisely because of a higher expected cost from a passed ball. Even the way in which a cricket batter handles the bat indicates a response to the greater cost of a passed ball. The typical cricket batter separates his or her hands, choking up with one hand while holding the other at the end of the bat. This allows him or her to more quickly and deftly impede a ball from passing to the wicket than does the hand-over-hand approach used in baseball.

The cricket grip favors defense over power. Indeed, power also carries less benefit in cricket than in baseball. To hit the ball out, or score a "sixer," in cricket, one typically needs to achieve a distance of roughly 250 ft (i.e., usually a bit more than the radius of a circular field with diameter approximately 500 ft). As the cricket batter does not stand in the exact center of the field, the exact distance needed to achieve a sixer depends upon where the ball is hit. In baseball, the batter must hit the ball into a 90-degree arc of a circle. To hit the ball "out" for a home run in *Major League Baseball*, the baseball batter typically needs to achieve a distance in excess of 400 ft. Given these stark differences in field dimensions, to adopt a baseball power grip in cricket would usually constitute an act of overkill.

There are additional reasons that baseball incentivizes a more offensive grip than does cricket. On a 360-degree cricket field with just 11 fielders, a cricket batter can earn a run from a well-placed, short hit. There are simply not enough fielders to defend all areas near the batter. However, baseball employs just two fewer fielders over a 90-degree field, a fact that allows baseball fielders to smother the infield and seldom give up an infield hit. The reader might counter this point by stating that *Major League Baseball* fields feature a greater "out distance" than do top-level cricket fields. Remember, however, that the batter is standing toward the middle of the field in cricket but on a border of the field in baseball. Despite its shorter "out distance," top-level cricket actually features a larger field than does baseball and even a larger field *per fielder*. The average *Major League Baseball* field is essentially a 90-degree arc of an imaginary circle with radius 400 ft, represented as the shaded portion of the circle in Fig. 5.1.

The average top-level cricket field is a full circle with radius 250 ft. That is, it is the full inner-circle in Fig. 5.2.

Fig. 5.1 A major league
baseball field as a 90-degree
arc of a circle (author-
rendered figure)

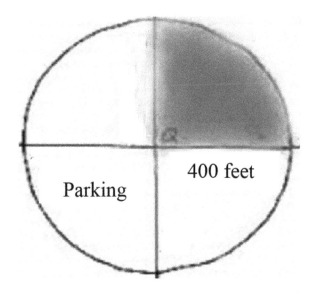

400 feet

Parking

Fig. 5.2 A top-level cricket
field as the inner circle of the
figure

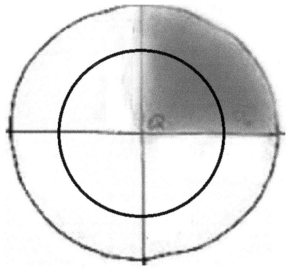

It is fairly evident that the area of the inner circle is larger than the shaded arc of the outer circle. To verify this, simply compare the magnitude of the white area in the inner circle to the magnitude of the shaded area not in the inner circle. It should be clear that the former area is greater than the latter. For still more verification, we can calculate the area of a 90-degree arc of a circle, where the circle's radius is 400 ft. This area is equal to $\frac{1}{4} \cdot \pi \cdot r_1^2$ or $\frac{1}{4} \cdot \pi \cdot 400^2$ square feet. This is equal to about 125,663 square feet. On the other hand, the area of a circle of radius 250 feet equals $\pi \cdot r_2^2$ or $\pi \cdot 250^2$. This is equal to about 196,349 square feet, substantially greater than the area of a standard *MLB* baseball field. Then, a top-level cricket field contains about (196,349 / 11) or roughly 17,850 square feet per fielder. In cricket, each fielder is in charge of the area of a large mansion. In *MLB*, the field contains about (125,663 / 9) or roughly 13,963 square feet per fielder. While this remains the area of a large mansion, the mansion that this area equates to is only about 78% as large as the previous. It is clear, then, that it is much more difficult to prevent short hits in cricket. It is simply a matter of geometry.

Many types of successful hits in cricket are likely to go as routine infield outs or foul balls in baseball. Therefore, the benefits of cocking the bat and gripping it hand-over-hand are greater in baseball than in cricket in that the former game allows relatively few opportunities to hit the ball safely over a short distance. Thus, we can invoke both cost and benefit differences to explain the contrasts between the respective stances of a cricket batter and baseball batter.

That's all I have to say about cricket versus baseball. I will leave the debate as to which game should be abolished on grounds of obvious inferiority to others.

References

1. Landsburg, S. E. (2007). *The Armchair economist: economics & everyday life.* Simon and Schuster.
2. Becker, G. S., & Stigler, G. J. (1974). Law enforcement, malfeasance, and compensation of enforcers. *The Journal of Legal Studies, 3*(1), 1–18.
3. Chang, Y. M., & Sanders, S. D. (2009). Corruption on the court: The causes and social consequences of point-shaving in NCAA basketball. *Review of Law & Economics, 5*(1), 269–291.
4. Coats, R. M., Pecquet, G., & Sanders, S. D. (2010). A fallacy in the ANWR drilling debate: A lesson on scarcity rents and intertemporal pricing under different market structures. *Journal of Industrial Organization Education, 4*(1), 1–12.
5. Hotelling, H. (1931). The economics of exhaustible resources. *Journal of Political Economy, 39* (2), 137–175.
6. United States Department of Agriculture. (2009). Factors behind the rise in global rice prices in 2008. Retrieved from https://www.ers.usda.gov/publications/pub-details/?pubid=38490
7. Hancock, J. (2008). Rice rationing at Sam's Club. Retrieved from https://www.baltimoresun. com/bs-mtblog-2008-04-evidence_of_rice_hoarding_from-story.html
8. The Economist. (2020). Marriage and divorce rankings. Retrieved from https://worldinfigures. com/rankings/index/218
9. Vrooman, J. (1995). A general theory of professional sports leagues. *Southern Economic Journal*, 971–990.
10. Schonbrun, Z. (2016). The batter's box gets a little boring. *New York Times.*. Retrieved from https://www.nytimes.com/2016/07/24/sports/baseball/the-batters-box-gets-a-little-boring.html.

Contest and Conflict in Society

6

Each person finds various ways of allocating resources toward the promotion of subjective well-being. This is not done in a vacuum but rather through a tenuous balance of cooperation and conflict with others. The following chapter examines the latter element of our interdependent experience.

The Theory of Relative Income

In 1949, economist James Duesenberry wrote a seminal paper in which he proposed the relative income hypothesis [1]. Subsequently, several important studies in microeconomics have uncovered strong evidence in favor of the hypothesis as a model of consumer behavior [2]. Despite the import of these findings, the hypothesis receives little treatment in contemporary economics texts. Of this disappearing act, leading economist Robert Frank writes [3],

> ...his (Duesenberry's) theory of consumer behavior clearly outperforms the alternative theories that displaced it in the 1950's—a striking reversal of the usual pattern in which theories are displaced by alternatives that better explain the evidence. His disappearance from modern economics textbooks is an intriguing tale in the sociology of knowledge. But it also has important practical implications. Unless we understand what drives consumption, which makes up two-thirds of total economic activity, we cannot predict how people will respond to policy changes like tax cuts or Social Security privatization...Most economists, it appears, just never wanted to believe the relative-income hypothesis—perhaps because it suggests the possibility of wasteful spending races (pp. 1, 2–3).

According to the relative income hypothesis, consumer choice depends upon income, prices, *and* community consumption standards. The last consideration, that consumer choice is influenced by community consumption, rests upon the premise that status is assigned to individuals based on consumption behaviors and that status, in turn, is determined within a community-specific context. For example, a $10,000 donation in a rural Mexican village might be looked upon differently than a $10,000

donation in Beverly Hills, California. Even though the magnitudes of these hypothetical expenditures are equal, their community-specific status implications are not.
Duesenberry's model departs from leading theories of consumer choice, which take
consumers as making decisions within a social vacuum. Within Duesenberry's
world, an individual's preferences and ultimate level of well-being depend upon
the consumption decisions and standards of others in her community.

In his original study on the subject, Duesenberry examined relative income effects
to better understand the relationship between income and savings. He found that the
rate of savings tends to increase with income in a cross-section or snapshot of US
households but that no such relationship is found in US households *across time*. That
is to say, a household with a current yearly income of $50,000 is expected to save at a
higher rate than a household with a current yearly income of $40,000. However, a
household with a current yearly income of $40,000 is not expected to save at a higher
rate if its income rises to $50,000 next year. Duesenberry showed that this apparent
paradox can be reconciled if one accepts relative income effects as being important to
consumer choice. One can think of the latter household as status-chasing the former
household. Given its position as status-chaser, the latter household is not expected to
save a lot of its income gains over time. In a status race, moreover, almost every
household has identified an aspirant group of households that it is chasing. These
effects lock households in status arms races such that they have difficulty translating
increasing income levels over time into higher rates of savings.

A household's participation in such status races is often referred to as "keeping up
with the Joneses." Let us assume that your neighboring family, the Joneses, owns
one 2008 *Toyota Camry*. This is fine and well, as your family also owns a 2008
Toyota Camry. One day, however, Mrs. Jones pulls up in a new *BMW 3 Series*.
Suddenly, your *Camry* looks a little worse for the wear, so you go out and get a new
BMW 3 Series. A few years later, Mrs. Jones upgrades to a *5 Series*, and you again
follow suit. At the end of the day, your family maintains an identical car as the
Joneses. However, it costs a whole lot more to maintain this equality of status than it
did in the days of *Camry*. Over time, your status race with the Joneses likely comes at
the expense of savings.

Interdependence of preferences implies that both absolute and relative income are
relevant to consumer choice such that your neighbor's purchase of a *BMW* might
influence both your well-being and your consumer behavior. That is, *their BMW*
purchase has potential implications upon *your* relative income level within the
community. Duesenberry writes [1],

> There is little observational warrant for the independence of different individuals'
> preferences yet it is implicit in most economic theory. The assumption has slipped in during
> the course of the historical development of consumer behavior theory...The interdepen-
> dence of preference systems has been recognized since the earliest days of economics. One
> can find discussions of emulation and the desire for distinction in the non-analytical parts of
> Jevons and Marshall, not to mention such writers as Veblen and Frank Knight (pp. 13–14).

Several studies demonstrate that relative income plays an important role in
determining a household's consumption behavior and overall well-being. Andrew

Clark and Andrew Oswald find that reported satisfaction levels over a group of British workers were negatively related to peer wage rates and comparison wage rates (i.e., wage rates of individuals to whom a worker compares herself) [4]. The authors explain,

> Despite what economics textbooks say, comparisons in the utility function seem to matter. This has a number of implications. In a world with comparisons, the case for growth as a way of increasing happiness is no longer so clear (see Easterlin, 1974 and Layard, 1980). Optimal tax policies are affected because there are negative externalities from high earners (see Oswald, 1983). In an analogous way, the wages offered by firms may have low variance if there are intra-firm comparison effects, and may rise over time if workers compare their current wage to their own previous wages (see Frank and Hutchens, 1993). Moreover, because preferences are intrinsically interdependent, the standard optimality results of the free market may fail to hold (p. 375).

A related micro-behavioral result demonstrates that relative income considerations greatly influence the outcome of ultimatum games in experimental studies. An ultimatum game is one in which there are two players who compete *and* cooperate with one another for a prize. We can assume, for the purpose of illustration, that the prize is $10. The first player is told to develop a sharing rule for the $10 prize. She might say, for example, "I keep $6, and you get $4." If the other player agrees to her ultimatum, the sharing rule applies. Otherwise, both players receive $0.

Standard economic theory predicts that the second player should accept any offer greater than $0. After all, even a penny is better than nothing. In practice, however, an ultimatum is often rejected if it piques feelings of unfairness in the second player. Houser and McCabe summarize ultimatum game experimental findings in a 2009 article [5].

> The key result of ultimatum experiments is that most proposers offer between 40 and 50% of the endowed amount, and that this split is almost always accepted by responders. When the proposal falls to 20% of the endowment it is rejected about half of the time, and rejection rates increase as the proposal falls to 10% and lower. As discussed by Camerer (2003) Chapter 2, ultimatum game results are highly robust to a variety of natural design manipulations (e.g., repetition, stake size, degree of anonymity, and a variety of demographic variables) (p. 47).

In other words, many second players are willing to forego a potential gain in order to be on an equal footing with the opposing player. The ultimatum game clearly reveals that relative payoffs matter in practice. In a study of the ultimatum game, Kagel, Kim, and Moser observe [6], "There is ample evidence that relative income shares entered players' utility functions, resulting in predictable variations in both rejection rates and offers" (p. 100). With the assumption of independent preferences, on the other hand, traditional economic theory does a markedly poor job in predicting the experimental outcomes of ultimatum games. Under the traditional assumption, it is as if experimental subjects are simply choosing to leave money on the table.

In a 2011 laboratory economic experimental study [7], economist Damian Damianov and I find that individuals are susceptible to entering costly status races

such that more income does not necessarily improve their well-being. We further
find that the pain associated with status positional spending races could be largely
avoided if incomes were publicly disclosed. Public income disclosure precludes the
value of such races as a means to signal income status. What, after all, is the value of
signaling something that is already known precisely? Under the 1967 *US Freedom of
Information Act*, the salaries of public employees in the United States must be
disclosed if requested. As a result, several websites in the United States have set
up searchable public employee salary databases. All else being equal, then, we
expect that US public employees tend to engage in less status-signaling behavior
than do their counterparts in the private sector.

It is clear from the literature that the relative income hypothesis holds substantial
empirical credibility, as well as a rich set of potential implications. If an economist
tries to tell you that more is always better, you should respond by inquiring, "For
whom?"

Conserving and Correcting for Status

As the previous essay suggests, positional considerations can lead us quickly into a
zero-sum status arms race. No matter how much we spend to signal our standing, the
total level of status conferred by a community remains essentially fixed. As a status
race ramps up, then, individuals spend more to secure status; yet average status
levels in the community remain the same. Several researchers call this phenomenon
the "status treadmill" or "hedonic treadmill" [8–10]. Consumers put forth substantial
resources to signal status but end up in the same place, on average.

Status races have characteristics that are similar to a game I used to play with
some of my junior high school classmates during our free period. I forget what we
called the game, but let us presently call it Stupidball because it was stupid and
involved a ball. In substance, Stupidball was some contrived mix of basketball and
football. For the most part, it was like basketball but with the added twist that the ball
handler could be tackled (onto the hard surface of the gym floor). At the beginning of
free period each day, I felt compelled to play a little Stupidball. At the end of free
period each day, I felt bruised and dizzy. Seeing me after a particularly violent
match, a friend of mine asked what had happened to me. I told him that I had been
playing Stupidball, or whatever I called the "game" at that time. My friend looked at
me in dismay. The next day, at the same time, my friend saw me in a similar
condition. A meaningful dialogue ensued between me and my friend. Here is the
gist of the transcript.

Friend: Have you been playing Stupidball again, Shane?
Me: Yes I have, friend.
Friend: Why do you even play that game?
[I consider the question and cannot manage to establish a reason. It hadn't occurred to me
    until that point that I even had a choice.]
Me: I guess I just see other people playing Stupidball and want to try to win a game.
Friend: Have you ever won at Stupidball?
Me: No.

Friend: Have you ever had fun playing Stupidball?
Me: No.
Friend: Then I guess you really shouldn't be playing that game.
Me: You're probably right.

My wise friend analyzed the game so well for me that I never played Stupidball again. I had not been playing the game because I liked it or because it represented a good skill to develop. Rather, I had been playing the game simply because it was some arbitrary competition that I could play to potentially win. Similarly, many people buy ridiculous status signaling goods not because they like them particularly or because the goods will directly help them in any way. They buy them to enter an arbitrary status competition that they believe to be winnable to some degree. Given their zero-sum nature, however, status competitions are much better at overheating credit card balances than at conferring winners. This is not to say that everyone who buys a *BMW* or a country club membership does so solely based on status considerations. Some people truly like the performance of the "ultimate driving machine" or the challenging golf course that a country club membership offers them. When a good is not attainable to all, however, it also becomes a means to signal one's status.

Status races are problematic, of course, in that each status expenditure by a given individual or household imposes an external cost—cost imposed on a party who is external to the transaction—on other members of the community. Moreover, there is an over-exchange of externally costly goods precisely because the buyer does not bear the full social cost of these purchases. For example, when Arthur buys an $80,000 sports car, he pays the sticker price of the car but does not pay the value of the status loss that he has imposed upon the rest of the community. Arthur and others making status purchases will naturally ignore the external cost of these purchases. As a result, there will be some status purchases for which the marginal private benefit to the buyer is less than the marginal social cost. Such purchases actually make society worse off, on net, and therefore account for the over-exchange problem associated with externally costly goods.

As an example, let us assume that Arthur derives $5000 of surplus (value above cost) from the sports car purchase. Based on his surplus amount and the price of the car, can you guess his value for the car? The answer is available in the following endnote.[1] Let us further assume that the car's seller derives a surplus of $5000 from the sale of the car. Based on the seller's surplus, can you guess the marginal cost of producing and distributing the car? This answer is also available in the following endnote.[2] As such, the total surplus or value created from the exchange is

---

[1] Arthur's value for the sports car is $85,000. A buyer's valuation, also known as her highest willingness to pay, for a unit of a good is equal to the good's unit price plus the buyer's surplus. Equivalently, the buyer's surplus equals her willingness to pay minus the unit price of the good.

[2] The marginal cost of producing and distributing the sports car is $75,000. A seller's valuation, also known as her *marginal cost* or *lowest willingness to accept* the exchange, is equal to the good's unit price minus the seller's surplus. Equivalently, the seller's surplus equals her marginal cost plus the unit price of the good. Seller's surplus is sometimes referred to as marginal profit. It is the additional profit that the seller earns from the sale of a unit of a good.

$5000 + $5000 = $10,000. In a market for a good that does not bear externalities, this would be the end of the story. The transaction would create $10,000 of net surplus for society. When markets function smoothly, they are, indeed, efficient in creating value by directing goods to those who value them most.

However, the story does not end there. Arthur did not buy a *Camry*, after all. He bought a nice sports car and has imposed an external cost upon his community of fellow status seekers. Let us say that there are 1000 people in Arthur's community and that the purchase of his sports car imposes an average external cost upon his fellow community members of $15. In other words, community members would pay up to $15, on average, to stop Arthur from purchasing the car. In aggregate, Arthur's sports car has imposed a $15,000 external cost upon the community. The total market surplus of the purchase was only $10,000. All told, society is actually $5000 worse off from Arthur's purchase. Such status purchases are not in the best interest of society such that there is an over-exchange of externally costly status goods in the world. The astute reader might ask why Arthur's fellow community members don't simply pool money and pay Arthur to not buy the car. Collectively, they would only have to pay Arthur $5000.01, or nominally more than his consumer surplus, to coerce him to forego the purchase. By doing so, others in the community would be better off by about $10,000.

Despite the obvious appeal of accumulating this coercion fund, other members of the community would have to (a) know of Arthur's plans in advance, (b) verify that Arthur isn't bluffing about his plans, c) estimate his consumer surplus from the purchase, and d) bear coordination costs in organizing against the purchase. If (a) Arthur does not speak about his intentions to buy the sports car, (b) Arthur is believed to be bluffing about the purchase, (c) it is difficult for the group to estimate his consumer surplus, or (d) the coordination costs to the group (e.g., time and effort spent organizing and accounting for contributed funds) are greater than $10 per other community member, then there will be no resistance to Arthur's purchase. In settings of collective choice, outcomes are often governed by factors such as asymmetric information and group coordination costs.

The theory of public choice, a field at the intersection of economics and political science, finds that such factors can have a profound influence upon collective choices in political elections and legislative voting. The field was founded largely by the late Nobel Laureate James Buchanan and the late political economist and legal scholar Gordon Tullock. But for his statements about the Nobel selection committee, many economists believe that Tullock would have shared in Buchanan's Nobel glory [11]. In several seminal papers and books, much of them completed in collaboration with one another, Buchanan and Tullock created a field that was not quite economics and not quite political science. Rather, public choice draws upon methodologies and subject matter from both disciplines. Their joint 1962 book, The Calculus of Consent [12], laid much of the foundation for the field. The theory of public choice is similar in content to social choice theory, which was the brainchild of another late Nobel

Laureate named Kenneth Arrow. Arrow created the theory of social choice in his 1951 doctoral dissertation. Arrow's work was later published as a book titled, Social Choice and Individual Values [13], which served to launch an active and important field of study. Not bad for a dissertation. In both the theory of public choice and social choice, it is found that our collective choices can be sub-optimal even if all individuals in a society act rationally. To quote a song written by Mick Jagger, a *London School of Economics* graduate, and fellow Stone Keith Richards, "You can't always get what you want."[3] Especially in a group setting.

When Two Wrongs Make a Right

Given the previous results, it is tempting to believe that status purchases are always bad for society. Sometimes in economics, however, two carefully combined wrongs do make a right. If centered around a particular type of good, status races may actually be useful. Let us motivate this idea with a bit more background concerning externalities. Whereas status goods are externally costly and therefore over-allocated from the perspective of society, environmentally friendly goods are externally beneficial and therefore under-allocated from the perspective of society. To understand why externally beneficial goods are under-allocated, consider what happens when you buy solar panels for the roof of your house. This purchase benefits your family in that it makes the air around you a little cleaner by displacing dirtier forms of energy production such as the burning of coal. However, it also benefits your neighbors, as they breathe the same general air as does your family. This latter benefit is an external benefit because the neighbors are external to your solar investment activity. Considering only the private benefits of solar panel purchases, then, not enough people will buy them from the perspective of society.

Whereas the purchase of a sports car imposes an external cost on your neighbors, the solar panel purchase externally *benefits* your neighbors. As such, any status race that forms around an environmentally friendly good has the potential to correct the under-allocation problem of such goods. Meanwhile, the environmentally friendly nature of the good will serve to correct the over-allocation problem created by the status race. Ideally, the two externality-based "wrongs" in this case will balance each other and create a good that is allocated with something like efficiency. The promotion of environmentally friendly consumption is, of course, no trivial matter. Air pollutants are among the most economically significant and pervasive examples of external cost. Air-bound externalities manifest themselves in the form of elevated rates of respiratory health problems, increased rates of infant mortality, increased rates of child and adult mortality, and rising average global temperatures [14–16]. Given the issue's supranational orientation, it is unclear that government

---

[3]Multiple songs written or co-written by Mick Jagger are, in fact, economics-themed. These include "You Can't Always Get What You Want" and "(I Can't Get No) Satisfaction," both of which have utility-themed lyrics.

regulation can offer a stand-alone solution. Moreover, the global pervasiveness of air-bound externalities limits the effectiveness of a single governmental actor in reducing greenhouse gas emissions. The late Nobel Laureate Elinor Ostrom brings up this point when calling for a polycentric approach to curbing greenhouse gas emissions [17].

Environmental regulation may be more than a simple coordination problem among nations. Nations may actually be in conflict with one another regarding environmental regulation. A game-theoretic study that I conducted with Bhavneet Walia and Abhinav Alakshendra [18] demonstrates that tighter environmental regulations by one country may actually provide incentives for other countries to loosen their own standards. When the first country tightens its standards, this raises production costs for many companies operating in that country. Responding to this increase in production expense, some such companies will likely seek to shed costs by relocating production facilities to a "pollution haven" country.

The story does not end there, however. Anticipating this offshore movement by firms in the first country, certain manufacturing-based, developing economies may actually lower their environmental standards to compete for the new offshore business. This last consequence is sometimes referred to as the "race to the bottom" [19]. As the name suggests, countries participating in the race to the bottom seek to achieve the lowest environmental regulations so as to attract greater levels of foreign direct investment. The following quote from the online magazine *The Broker* states [20],

> The debate about the 'race to the bottom' hypothesis focuses mainly on globalization and the entry of developing countries into the global market. The idea is that international trade and investment will turn to lower cost countries more easily when these countries become more integrated in the world economy. To attract investment, countries open their markets but due to international competition they are tempted to relax labour, environmental and tax regulations, at the detriment of social policies, and resulting in a race to the bottom.

Unlike national environmental regulations, status races that involve environmentally friendly goods have the potential to scale productively at a global level. Some such goods have already spurred global status races. There is evidence, for example, that consumers pay a premium for the *Toyota Prius* due to its ability to signal one's status as an environmentally friendly consumer. A study by Delgado, Harriger, and Khanna finds that *Prius* owners pay a fairly substantial premium for the car's environmental signaling value [21]. In the study, the authors compare the price distribution of *Prius* cars relative to that of other hybrid cars that cannot be distinguished from their "non-hybrid counterparts." For example, the *Toyota Camry* hybrid is basically indistinguishable from the non-hybrid *Camry*. Therefore, the hybrid *Camry* does not signal environmental consciousness to the same extent that the *Prius* does. By pricing the respective components of each car, the authors estimate that *Prius* owners pay a $587 premium to signal environmental friendliness. In the case of the *Prius*, at least, an environmental friendliness signal is estimated to cost about as much as a mid-tier laptop computer or a haircut from famed stylist Chris McMillan.

Using market-level vehicle ownership data, a related article by Sexton and Sexton finds that peer effects increase *Toyota Prius* demand in the US states of Colorado and Washington [22]. The authors attribute this result to "positional conservation," whereby they mean to indicate the demonstration of environmental friendliness for the purpose of garnering status. Sexton and Sexton suggest that positional conservation is a relatively new dimension of status signaling that can sometimes result in seemingly ridiculous behavior. They state, "...status conferred upon demonstration of environmental friendliness is sufficiently prized that homeowners are known to install solar panels on the shaded sides of houses so that their costly investments are visible from the street." In such cases, homeowners are actually willing to *be* less environment friendly in order to *appear* more environmentally friendly. Studies by Schultz et al. [23] and Alcott [24] also find that people are friendlier to the environment when their actions are observed. Specifically, they show that household energy consumption is expected to decrease significantly when tracked and judged against that of neighboring households. In the book Nudge, Nobel Laureate Richard Thaler and co-author Cass Sustein provide an in-depth discussion as to the positionality of environmentally friendly goods [25].

As mentioned previously, the basic requirements of a positional good are that it be both visible and not accessible to everyone. This usually means that the good is ostentatious and expensive. Some positional goods, however, are inaccessible not due to their price but rather because they entail some nonmonetary sacrifice that much of the population is unwilling or unable to make. The *Prius*, for example, is a moderately affordable car. A 2020 *Prius* lists for $24,325, and some US states offer generous rebates on hybrid car purchases. California, for example, offers up to $7000 in subsidies toward the purchase or lease of a hybrid car. What, then, makes the *Prius* inaccessible to some? It is partly the modest size of the *Prius* that creates the potential for sacrifice on the part of the owner. The car cannot haul large loads or even fit the entirety of an oversized family at one time. Moreover, the *Prius* certainly won't intimidate other drivers on the road (in case that's your thing). For the *Prius* to work as a family car, the family must be willing to make certain sacrifices. By and large, only an environmentally friendly family would be willing to make these sacrifices. Hence, the *Prius* bestows status not so much via its price tag but, rather, via what it signals about the owner's preferences.

As tempting as it may be, we should not blame the goods that we use to signal status for the existence of status races. It is not the *Tiffany* diamond or the *BMW i8 Roadster* that causes status races. Rather, the cause of status races is within each of us. Particularly, it is in our dorsal anterior cingulate cortices, our parietal cortex, our ventral striatum, and other regions of the brain that regulate social comparisons and rewards therefrom [26]. It is our deep-seated biological drive for status that causes status races. If *Tiffany*, *BMW*, and all other luxury brands ceased to exist overnight, humankind would regroup and form status races around a new set of brands and goods. Positional goods are simply the pawns in our status race games. Robert Frank defines them [27] as those goods "whose value depends relatively strongly on how they compare with things owned by others," whereas the value of non-positional goods depends "relatively less strongly on such comparisons..." (p. 101). From

Thorstein Veblen [28] to James Duesenberry [1] to Robert Frank [29, 30] and Erzo Luttmer [31], an undercurrent of the economics literature has established a broad class of goods that are purchased at least in part to signal status [4, 32–36]. This goods class is broad and includes houses, vacations, private school enrollments, designer suits, and dinner reservations at *Michelin*-starred restaurants.

Earlier in the chapter, we established that status races can decrease the rate at which individuals save. While a decrease in the savings rate can lead to problems, it does not imply that individuals are worse off, *per se*. However, several economic studies find evidence that status races do, in fact, lead to large welfare losses via the imposition of external costs [31, 37, 38]. To understand this outcome, let us reconsider the scenario in which you and the neighboring Jones family each own a 2008 *Toyota Camry*. At that moment in time, you and the Joneses are equals in terms of status conferred by quality of car. One day, the Joneses buy a *BMW 3-Series* car. Mrs. Jones drives the car back from the dealership and, though it still possesses that showroom shine, immediately begins to wash this specimen of a new car at the edge of her driveway. You observe this scene and are instantly taken aback: *How ostentatious can you get? How crass? Is that the 335i or the 335xi?* Completely miffed, you find your binoculars, confirm that the Jones family now owns a *335xi*, and head straight over to the *BMW* dealership. A few hours later, you are standing in your driveway washing your very own showroom clean *335xi*. Across the street, Ms. Jones is now rummaging around her utility closet for her binoculars. Suddenly, she doesn't feel so special.

Having each "graduated" from old *Camrys* to new *BMWs*, it appears that both families will be better off as a result. Such an outcome is likely to obtain only if you and the Joneses each bought a *BMW* for non-positional reasons (e.g., for its performance and reliability). Some aspects of this story lead us to second-guess the factors motivating these purchases, however. It is curious, for example, that both you and the Joneses started with the same car in the first place. This may indicate that you and the Joneses have been countering the positional purchases of one another for a long period of time heretofore. It is further curious that you purchased a *BMW 3-Series* right after the Joneses made an identical purchase. Given these circumstances, it appears as though you and the Joneses are locked in a positional spending race. As a result of the race's intensification, both families have less savings and less disposable income to spend on non-positional goods. Despite all of this sacrifice, you still haven't gained any car ownership related status edge on the Joneses and vice versa. While this example may seem a heavily stylized one, there is a great deal of evidence that humans are, in fact, very herd-like in their positional consumption behavior [39].

In an important study of positional spending, leading economist Robert Frank constructs a theoretical model of strategic consumer choice. Frank finds that positional spending contests lead to large social welfare losses [27]. This result is driven by overconsumption of positional goods. In a subsequent study, Frank observes [30], "Recent years have seen renewed interest in economic models in which individual utility depends not only on absolute consumption, but also on relative consumption. In contrast to traditional models, these models identify a fundamental conflict between individual and social welfare" (p. 137). Hopkins and Kornienko construct a status game and find that, while this conflict exists generally, the degree of conflict depends on the level of income inequality in a society [40]. They write, "In the symmetric Nash equilibrium, each individual spends an inefficiently high amount on the status good...(Status) signaling is costly and the Nash equilibrium is Pareto dominated by the state where agents take no account of status in their consumption decisions" (pp. 1085, 1099). By "symmetric Nash equilibrium," Hopkins and Kornienko mean to indicate the set of expected status expenditures when all individuals have the same income level. When incomes are the same or similar, we can think of individuals and households as having trouble separating from the pack in terms of status expenditures. Therefore, status expenditures can be intense and costly in such cases.

In general, a Nash equilibrium is simply a solution to a strategic interaction in which each individual has chosen a best response to the other individual(s). As each individual has selected a best response, there are no gains from unilateral deviation in a Nash equilibrium. As in the case of any equilibrium concept, then, we expect a Nash equilibrium to persist once reached, all other things being equal. Equilibria are what we might call stable or self-perpetuating. In the present case, a Nash equilibrium would be a state in which each consumer has chosen a best response level of status good expenditure given the status expenditures of others in the community. Of course, the Nash equilibrium concept was developed by its famous namesake, John Nash. As Kenneth Arrow did a year later, Nash contributed this Nobel Prize-winning concept in the form of a doctoral dissertation.

In a related study, Fershtman and Weiss [41] develop the notion of "social rewards" and find that positional spending need not always cause conflict between individual and social welfare. In fact, positional spending can lead to gains in social welfare if the positional good of interest would be under-consumed in the absence of positional considerations. They state, "If it is desirable to reduce output (of an action), because the output causes pollution for instance, this must be done by other means such as legal enforcement. Social rewards will be effective only if it is desirable to increase output, because of positive externalities". This finding suggests that the exchange of externally beneficial goods, which are under-exchanged in the absence of market intervention, might be achieved if social (status) rewards are attached to the consumption of the externally beneficial good.

Higher education is externally beneficial in that human capital formation creates knowledge spillovers. Those who go to college will go on to benefit not only themselves but also their co-workers with the knowledge and skills they gain. As such, we expect that higher education is under-consumed in countries that do not offer sufficient tuition subsidies. This under-consumption can be corrected if the pursuit of higher education bestows additional status upon graduates. In fact, we do observe that higher education often exhibits positionality. In a 2014 article, Botha finds that educational attainment in South Africa exhibits properties consistent with those of a positional good [42]. In some ways, this result is intuitive. It is something like a basic truth that graduation from an elite university confers instant social status upon the graduate. Whereas a *State U.* graduation typically feels something like a July 4th barbecue, an *Ivy League* graduation is more akin in spirit to a coronation. To graduates of *State U.*, such as myself, the words "*Harvard* graduate" can ring in the ears for some time, especially when spoken by the artifact herself. If this ringing reaches the inner ear, it can sometimes lead to dizziness and a general loss of equilibrium.

In a 1926 short story titled "The Rich Boy" [43], F. Scott Fitzgerald wrote, "Let me tell you about the very rich. They are different from you and me." Fitzgerald had previously spoken this line to fellow writer Ernest Hemingway, to which Hemingway replied, "Yes, they have more money" [44]. Despite Hemingway's apt reply, Fitzgerald had a point that relates strongly to the status implications of elite education. Namely, some feel that the *Ivy League* educated are different from you and me. To which Hemingway might have said, "Yes, they have more test prep training." The status implications of higher education almost certainly do not stop at the level of elite universities. For example, even mid-tier universities may confer social status upon graduates who return to communities with relatively low rates of higher educational attainment. (*He's the smart one. He went to college.*)

In a study of higher education as a positional good, Fershtman, Murphy, and Weiss discuss the corrective powers of status races when framed around educational attainment [45]. Rather than lead us into a social welfare destructive form of status competition, status races focusing on educational attainment have the ability to motivate an externally beneficial activity that is otherwise under-subscribed within a free-market setting. As with educational attainment, status races involving environmentally friendly goods have the potential to confer much-needed external benefits on a global scale. We might conclude, then, that it is not the status race but what we center the status race around that determines how we fare as a society. In the end, we are slave to our underlying values.

Real-Life Problems

When a person comes to appreciate the fictional world presented in a particular television sitcom, she may become tempted to believe that the actors are exactly the same in real life as the imperfect characters that they play. Rather than truth, this notion usually turns out to be only a symptom of one's naïve attempt to live vicariously through a fictional storyline. You see, the actors in a television show are often more imperfect, and less redeemable, than their fictional counterparts. This is true, of course, because actors are faced with real problems, while fictional characters deal with fake problems that are destined to resolve themselves in one way or another.

Seinfeld, the popular television comedy of the 1990s, is a strong case in point. While the characters held together in their dysfunctional way, the real-life actors have not always demonstrated such unity. The chief elements of the group's undoing seem to be money and ego. While Jerry Seinfeld the fictional character always allowed Kramer to loot his refrigerator for food, albeit reluctantly, Seinfeld the person has a much harder time sharing. In the years following the show's end, he resisted pressure from his three co-stars to share in syndication profits, as well as in profits from the sale of the show's DVD [46]. The latter refusal caused two of Seinfeld's co-stars, Julia Louis-Dreyfus and Jason Alexander, to forego participation in interviews for the DVD. In a media interview about this state of affairs, Jason Alexander stated [46], "Julia, Michael and I, during our big renegotiation for the final year, asked for something that I will go to my grave saying we should have had, and that is back-end participation in the profits for the show."

Upon hearing about this sad state of affairs, my first thought was not that Person A deserves this or Person B deserves that. My first thought was that, in terms merely of profitability, it was not smart for Jerry Seinfeld and co-creator Larry David to offer the show's co-stars nothing more than a standard studio fee in exchange for their interviews. To some degree, the DVD is an inferior product without interviews from Dreyfus and Alexander. In other words, the demand curve for *Seinfeld* DVDs would have shifted outward if interviews from Dreyfus and Alexander were included. Other things being equal, this demand shift would have led to greater revenues from the sale of the DVD. Even if including the interviews produced only a tiny demand shift, in relation to the magnitude of sales, it remains that Seinfeld and David could have improved the situation for all parties involved. They could have offered the co-stars some amount of "back-end" remuneration such that (a) the co-stars would receive reciprocity beyond the standard studio fee and (b) their own profit from the sale of the DVD would *increase*, on net, in expectation. In addition, consumer surplus, or the total value consumers derive from purchasing the DVD beyond the amount paid, would have risen with the inclusion of the interviews. When economic value of any kind is created, it is like forming a new pie of some size. As with physical pies, this figurative pie can typically be divided to everyone's benefit.

To summarize, by negotiating with the show's co-stars, Seinfeld and David could have increased the girth of their own respective wallets and those of the show's co-stars, while simultaneously increasing the degree of consumer well-being created by the DVD. After considering these facts, I came to two tentative conclusions: (1) Seinfeld and David are probably better at comedy than business, and (2) the people handling the business affairs of Seinfeld and David are themselves not so good in business. Then, my friend enlightened me. Seinfeld and David, like most humans, value more than just money in the bank. They also value the sense that they are unique and creative. By conceding anything more than a standard actor's fee to the show's co-stars, they would have been saying something like, "You are more than standard actors in a television show. Creatively, you helped make *Seinfeld* what it is and deserve credit for that." Perhaps then, Seinfeld and David value protection of image more than any profits that can be garnered from improving the show's DVD (wherever that DVD has now come to rest in the present *Age of the Cloud*). Hence, it may not be so much a lack of business acumen but, rather, an agglomeration of ego that has denied all parties involved a form of closure that can come only from DVD bonus features.

The Limitations of Journalistic Objectivity

A core principle of journalistic writing is that the journalist should write from a place of *objectivity*. There is a simple method for testing whether a journalist has written a particular article objectively. Upon reading an article, the reader should try to guess whether the journalist's opinion on the topic of the article aligns with her own. If, based solely on passages of the story, the reader is fairly certain that the journalist's view on the topic aligns (or misaligns) with her own, then the journalist has probably not remained objective. If, on the other hand, the reader is uncertain how the journalist's view relates to her own, then the journalist has probably remained objective. This test works better if taken without knowledge of the journalist's name. Ideally, the reader would not know even the name of the newspaper or magazine in which the article appears before taking the test. However, such ignorance may be difficult to achieve, as readers typically select their periodicals in advance of reading an article.

In many journalism schools, the topic of objectivity is discussed as if having the precision of a scientific standard. However, there are many open questions surrounding this topic. Among these: *What does it mean to write objectively? Do newspapers produce news with absolute objectivity? Why do journalists aspire to be objective in the first place?*

To write objectively means to quarantine one's views on a topic when penning a piece. To adapt a line from fictional detective *Joe Friday* of *Dragnet* fame, an objective journalist provides "just the facts." While professional standards such as absolute objectivity are regularly preached at the highest levels of journalism, it is my contention that we do not receive absolute objectivity from any newspaper in the world. You see, implicit in each newswriter's choice of material and presentation is an innate *human* bias that cannot be quarantined from the written page. A clear example of human bias is observed in the old Dick Cheney hunting accident story [47]. Below, I construct an account of the story that somewhat approximates absolute objectivity.

> Members of Species A were trying to kill members of Species B. One member of Species A accidentally wounded another member of Species A. The efforts of Species A to kill Species B were abruptly halted, likely sparing multiple members of Species B.

Of course, previous news accounts of this incident were much more concerned with the status of Cheney's fallen hunting partner, Harry Whittington (i.e., the other member of Species A), who remains alive at 93 years old as of this writing. It is clear that newspapers cannot simultaneously reflect *human values* and maintain objectivity in the absolute sense. Indeed, newspapers are written *for humans*, as we are the only known species that has figured out how to read. A newspaper written from the perspective of giraffes, or one written in a manner that exercised neutrality between the human and giraffe perspectives, would almost certainly not be read by giraffes (unless giraffes are hiding something) and would probably not be read by humans either. The writer cannot write too broadly, or she will certainly miss her audience.

Therefore, it is probably not feasible for journalists to seek absolute objectivity. However, many newspapers attain some degree of objectivity conditional upon a set of basic human values. In other words, after taking for granted a set of values upon which most of us seem to agree (i.e., a quail death or two is less newsworthy than one person injury), such newspapers are basically objective. Many newspapers and other news sources diverge from this basic objectivity in different ways. These instances of idiosyncratic divergence give rise to the spectrum of ideologies, political and otherwise, that we observe among the set of available news sources.

However, why do journalists practice, even idealize, any degree of objectivity? After thinking about this question for some time, I reasoned that people have traditionally substituted between news media and old-fashioned word-of-mouth. After all, many people continue to receive news residually on the neighbor's front porch or around the office water cooler. In terms of profitability, then, it is in the interest of the news industry to differentiate its product from this informal competitor. The news industry does so by offering compactness and some degree of objectivity. The former attribute allows people to receive many different stories at the flip of a page rather than by talking to dozens of people. The latter allows people to receive news information without a lot of exaggeration or human error. In other words, people value objectivity to an extent because it allows them to interpret the news more consistently and thus with heightened meaning.

However, people seek at least one more thing when reading a newspaper: a reflection of their values in the choice and framing of stories. This is, of course, where objectivity must end. Hence, it is the reader who has the final word in establishing the nature and limitations of journalistic standards. There is an old business slogan that states, "The customer is always right." In the case of partisan news outlets specifically, this slogan is something of a tautology. That is, the consumer of a partisan news source usually chooses the source because its specific brand of ideology closely matches her own [48]. After all, we don't see a lot of Bernie Sanders supporters reading *National Review* or a lot of Alt-Right supporters setting slate.com as their homepage. After a partisan news choice is locked in, it becomes something of a reflection of the reader's intrinsic views. A feedback loop that tells the reader she is always right.

Blood Conflict

Part of my scholarly research involves the study of international conflict, but this fact does not make me a political scientist. More than a subject with a set body of material, economics is a scientific method by which decision-making is studied in all its various forms. Edward Lazear brings home this point quite well in a journal article titled, "Economic Imperialism" [1]. In the article, Lazear relates that economics is, first and foremost, a methodology for studying human behavior in a way that is unique from the other social sciences in its rigor and self-consistency.

The studies of conflict I've completed with Yang-Ming Chang, Bhavneet Walia, Joel Potter, James Boudreau, and others [49–52] employ the economic methodology. That is, we assume that countries act collectively in making rational or self-interest maximizing decisions. Upon this premise, we establish mathematical models of conflict in which two or more parties seek to maximize their well-being through conflict decisions. Most of the time, a country maximizes its well-being by acting peacefully toward another country. That is, peace between two countries is the most common equilibrium point. Even a country that is engaged in war with one other country has implicitly decided to act peacefully toward all other countries. Let us consider the number of country pairs, or dyads, among the 195 United Nations member countries in the world. Among a set of 195 countries, there are $C_2^{195} = \frac{195!}{2!193!}$ $= 18,915$ country pairs. If 19 of these pairs are engaged in international conflict with one another at a given point in time, then the world is 99.9% free of international conflict. That is, the modern world is usually about as effective in avoiding international conflict as hand sanitizer is in eradicating bacteria from a surface. While any level of conflict and violence in the world is difficult for many of us to accept, we have come a long way since the bloodbaths that were the twentieth, nineteenth, eighteenth, ..., centuries of human history. An excellent magazine article by Zack Beauchamp illustrates this trend over different time periods, demonstrating the degree to which humans have settled down in the late twentieth- and early twenty-first-century periods [53]. Using the *PRIO Battle Deaths* data set [54] and supplementary data on world population [55], I have rendered Fig. 6.1 to show the trend battle casualty rate by year (1946–2018).

Figure 6.1 demonstrates that the world battle death rate has trended downward throughout the post-WWII era. In the direct aftermath of WWII, this decrease was abrupt for clear reasons (i.e., the transition from having a world war to not having a world war). Subsequently, the decline has continued at a more gradual rate. To what do we owe this change? There are several possibilities. Following WWII, governing entities such as the United Nations and European Union were created to promote peace and unity through world governance and collective action. Moreover, the Cold War between the United States and U.S.S.R. may have had the ironic (net) effect of decreasing the global battle death rate. By stockpiling nuclear arms, these two entities inspired the threat of "mutually assured destruction," which can deter direct conflict from ever taking place. During the latter half of the twentieth century, moreover, the world experienced sustained economic growth. This may have deteriorated the economic incentives for conflict to some degree. Indeed, several of the most horrific conflicts that did occur during the late twentieth century took place in underdeveloped regions of Africa. Lastly, it is possible that infusions of democracy and female political leadership have shifted governance priorities in our modern world. However, such factors require additional study as they relate to conflict.

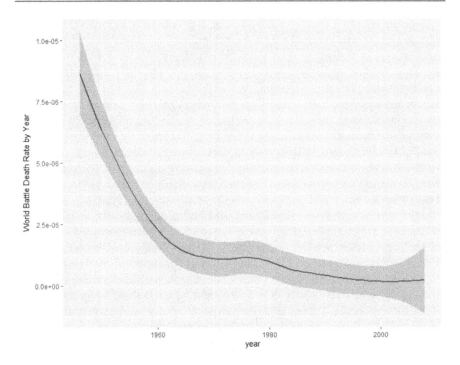

Fig. 6.1 Trend world battle casualty rate by year (1946–2018). Plot rendered in *R* using the *tidyverse* package. R code for plotting this available in Appendix. Source Data: *PRIO Battle Deaths* data set [54] and supplementary data on world population [55]. Interpolation of this data was used to estimate the world population for years 1946–1950

But Doesn't a Game Have to Be Fun?

Game-theoretic models of conflict can lend insights as to why armed conflicts occur between two countries. Further, a "dynamic" model of conflict (i.e., one extended over multiple periods) can tell us whether a conflict equilibrium will persist. Of course, the persistence of war is a vital issue to individuals in conflict-affected areas. Such individuals must often decide whether to hide, fight, or flee their homeland, and central to this decision is the issue of how long conflict will persist. In addition to the direct toll of warfare, which is sizable, persistent conflict creates a host of indirect social problems within the affected region. Extended lawlessness in a conflict region creates an ideal environment for opportunistic criminal behavior. Further, the extended loss of schooling to children acts to reinforce the relative attractiveness of criminal activity by suppressing the human capital of future generations within the state. Rather than merely a symptom of a failing state, persistent conflict is likely to contribute to a state's very failure.

There are several aspects of a conflict that may contribute to its level of persistence. A study that I conducted with Yang-Ming Chang and Joel Potter [56] suggests that conflict is more likely to persist in a sort of stalemate scenario when two parties are evenly matched in military terms. Imagine, for example, two people engaged in a wrestling match. Let these opposing parties be carbon copies of one another in strength and technique. As such, neither wrestler is likely able to execute a decisive move upon the other. These wrestlers are locked in a classic grudge match, whereby the only thing that can end the match is the clock. Unfortunately for humanity, there is no clock that regulates international conflicts. In such settings, however, alternative forms of regulation can be effective. For example, third-party intervention can be important in resolving conflict between two evenly matched parties before a stalemate is reached.

In the same study, Yang-Ming Chang, Joel Potter, and I also find evidence that the degree to which conflict destroys the very resources being contested influences the conflict's persistence. As an example, we can compare a "blood diamond conflict" with a conflict that decides control over a constructed oil pipeline. Whereas conflict can destroy part of the value of a constructed pipeline (e.g., through bombing), it cannot very easily destroy the value of minerals in the ground. For a conflict that renders destruction of the contested resource, we expect less persistence of fighting. That is, we expect that parties are increasingly likely to lose interest in conflict as the conflict prize evaporates. On the other hand, conflicts are more likely to persist if the prize value itself persists through conflict. As an alternative example of conflicts with persistent prize value, consider wars over religious lands. No matter how many bombs fall on the West Bank, it will retain its religious significance to both parties who contest its rights. From this perspective, it is not a great surprise that the West Bank has been fought over for most of the post-WWII era.

Conflict is a tragic product of human nature. With policies that take into account the calculus of conflict, however, we can act to minimize its toll upon humans. A common refrain states, "Can't we all just get along?" The economic view is that if in fact we can, it is often through policy measures that tip the balance toward a peaceful equilibrium.

References

1. Duesenberry, J. S. (1949). *Income, saving, and the theory of consumer behavior*. Cambridge: Harvard University Press.
2. Kosicki, G. (1987). The relative income hypothesis: A review of the cross section evidence. *Quarterly Journal of Business and Economics*, 65–80.
3. Frank, R. H. (2005). Does absolute income matter? In L. Bruni & P. L. Porta (Eds.), *Economics and happiness* (pp. 65–90). Oxford: Oxford University Press.
4. Clark, A. E., & Oswald, A. J. (1996). Satisfaction and comparison income. *Journal of Public Economics, 61*(3), 359–381.
5. Houser, D., & McCabe, K. (2009). Experimental neuroeconomics and non-cooperative games. In *Neuroeconomics* (pp. 47–62). Academic Press.
6. Kagel, J. H., Kim, C., & Moser, D. (1996). Fairness in ultimatum games with asymmetric information and asymmetric payoffs. *Games and Economic Behavior, 13*(1), 100–110.

7. Diamonov, D., & Sanders, S. (2011). Status spending races, cooperative consumption and voluntary public income disclosure. *International Review of Economic Education, 10*(1), 29–53.
8. Sanders, S. (2010). A model of the relative income hypothesis. *The Journal of Economic Education, 41*(3), 292–305.
9. Binswanger, M. (2006). Why does income growth fail to make us happier?: Searching for the treadmills behind the paradox of happiness. *The Journal of Socio-Economics, 35*(2), 366–381.
10. Firebaugh, G., & Schroeder, M. B. (2009). Does your neighbor's income affect your happiness? *American Journal of Sociology, 115*(3), 805–831.
11. Xue, Z. (2016). Fond memories of professor Gordon Tullock. *Journal of Bioeconomics, 18*(2), 113–114.
12. Buchanan, J. M., & Tullock, G. (1962). *The calculus of consent*. Ann Arbor: University of Michigan Press.
13. Arrow, K. J. (2012). *Social choice and individual values*. Yale University Press.
14. Chay, K., & Greenstone, M. (1999). *The impact of air pollution on infant mortality: Evidence from geographic variation in pollution shocks induced by a recession* (NBER Working Paper No. 7442).
15. Ruhm, C. (2000). Are recessions good for your health? *Quarterly Journal of Economics, 115* (2), 617–650.
16. Lashof, D. A., & Ahuja, D. R. (1990). Relative contributions of greenhouse gas emissions to global warming. *Nature, 344*, 529–531.
17. Ostrom, E. (2010). Beyond markets and states: Polycentric governance of complex economic Systems. *American Economic Review, 100*, 641–672.
18. Sanders, S., Alakshendra, A., & Walia, B. (2014). National emissions standards, pollution havens, and global greenhouse gas emissions. *American Journal of Economics and Sociology,* 73.2:353–368.
19. Konisky, D. M. (2007). Regulatory competition and environmental enforcement: Is there a race to the bottom? *American Journal of Political Science, 51*(4), 853–872.
20. Bieckmann, F., & Evert-Jan, Q. (2015). Retrieved from https://www.thebrokeronline.eu/ideals-versus-reality-d83/
21. Delgado, M. S., Harriger, J. L., & Khanna, N. (2015). The value of environmental status signaling. *Ecological Economics, 111*, 1–11.
22. Sexton, S. E., & Sexton, A. L. (2011). *Conspicuous conservation: The Prius effect and willingness to pay for environmental bona fides* (Working Paper). University of California at Berkley.
23. Schultz, P. W., Nolan, J. M., Cialdini, R. B., Goldstein, N. J., & Griskevicius, V. (2007). The constructive, destructive, and reconstructive power of social norms. *Psychological Science, 18*, 429–433.
24. Alcott, H. (2009). *Social norms and energy conservation* (Working Paper). Department of Economics, New York University.
25. Thaler, R. H., & Sunstein, C. R. (2009). *Nudge: Improving decisions about health, wealth, and happiness*. Penguin.
26. Kedia, G., Mussweiler, T., & Linden, D. E. (2014). Brain mechanisms of social comparison and their influence on the reward system. *Neuroreport, 25*(16), 1255.
27. Frank, R. H. (1985). The demand for unobservable and other nonpositional goods. *American Economic Review, 75*(1), 101–116.
28. Veblen, T. (1899). *The theory of the leisure class*. New York: The New American Library 1953.
29. Frank, R. H. (2008). Should public policy respond to positional externalities? *Journal of Public Economics, 92*(8–9), 1777–1786.
30. Frank, R. H. (2005). Positional externalities cause large and preventable welfare losses. *American Economic Review, 95*(2), 137–141.
31. Luttmer, E. (2005). Neighbors as negatives: Relative earnings and well-being. *Quarterly Journal of Economics, 120*, 963–1002.

32. Heffetz, O. (2011). A test of conspicuous consumption: Visibility and income elasticities. *The Review of Economics and Statistics, 93*(4), 1101–1117.
33. Kosicki, G. (1987). A test of the relative income hypothesis. *Southern Economic Journal, 54*, 422–434.
34. Clark, A. E., & Oswald, A. J. (1998). Comparison-concave utility and following behaviour in social and economic settings. *Journal of Public Economics, 70*, 133–155.
35. Kagel, J., Chung, K., & Moser, D. (1996). Fairness in ultimatum games with asymmetric information and asymmetric payoffs. *Games and Economic Behavior, 13*(1), 100–110.
36. Oswald, A. J. (1997). Happiness and economic performance. *Economic Journal, 107*, 1815–1831.
37. Easterlin, R. A. (1995). Will raising the incomes of all increase the happiness of all? *Journal of Economic Behavior and Organization, 27*, 35–47.
38. McBride, M. (2001). Relative-income effects on subjective well-being in the cross-section. *Journal of Economic Behavior & Organization, 45*(3), 251–278.
39. Chen, X., & Zhang, X. (2009). Blood for social status: Preliminary evidence from rural China. No. 319-2016-9850.
40. Hopkins, E., & Kornienko, T. (2004). Running to keep in the same place: Consumer choice as a game of status. *American Economic Review, 94*, 1085–1107.
41. Fershtman, C., & Weiss, Y. (1998). Social rewards, externalities and stable preferences. *Journal of Public Economics, 70*(1), 53–73.
42. Botha, F. (2014). Life satisfaction and education in South Africa: Investigating the role of attainment and the likelihood of education as a positional good. *Social Indicators Research, 118*(2), 555–578.
43. Fitzgerald, F. S. (1926). *The rich boy* (pp. 317–49). Feedbooks.
44. Johnston, D. C. (2005). Richest are leaving even the rich far behind. *The New York Times, 5*.
45. Fershtman, C., Murphy, K. M., & Weiss, Y. (1996). Social status, education, and growth. *Journal of Political Economy, 104*(1), 108–132.
46. Bolluyt, J. (2018). Here's how much money the cast of *Seinfeld* makes for reruns. Retrieved from https://www.cheatsheet.com/entertainment/what-the-seinfeld-cast-makes-for-reruns.html/
47. The following academic article discusses news coverage of the Dick Cheney hunting accident story: Theye, K. (2008). Shoot, I'm sorry: An examination of narrative functions and effectiveness within Dick Cheney's hunting accident apologia. *Southern Communication Journal, 73*(2), 160–177.
48. Iyengar, S., & Hahn, K. S. (2009). Red media, blue media: Evidence of ideological selectivity in media use. *Journal of Communication, 59*(1), 19–39.
49. Chang, Y. M., Potter, J., & Sanders, S. (2007). War and peace: third-party intervention in conflict. *European Journal of Political Economy, 23*(4), 954–974.
50. Chang, Y. M., & Sanders, S. (2009). Raising the cost of rebellion: the role of third-party intervention in intrastate conflict. *Defence and Peace Economics, 20*(3), 149–169.
51. Sanders, S., & Walia, B. (2014). Endogenous destruction in a model of armed conflict: Implications for conflict intensity, welfare, and third-party intervention. *Journal of Public Economic Theory, 16*(4), 606–619.
52. Boudreau, J. W., Sanders, S., & Shunda, N. (2019). The role of noise in alliance formation and collusion in conflicts. *Public Choice, 179*(3–4), 249–266.
53. Beauchamp, Z. (2015). Six hundred years of war and peace, in one amazing chart. Retrieved from https://www.vox.com/2015/6/23/8832311/war-casualties-600-years
54. Lacina, B., & Uriarte, G. (2009). The PRIO battle deaths dataset, 1946-2008, Version 3.0 Documentation of Coding Decisions. *CSCW/PRIO*.
55. Population data source: Worldometers. (2020). World population by year. Retrieved from https://www.worldometers.info/world-population/world-population-by-year/
56. Vrooman, J. (1995). A general theory of professional sports leagues. *Southern Economic Journal*, 971–990.

More on Politics and Policy

7

This chapter further considers the relevance of economic thought to public policy matters big and small.

When It Comes to Political Influence, Nothing Is Sacred...Not Even Religion

Religious interest groups permeate the US political landscape. This is why, for example, much of our country harkens back to the Prohibition Era each Sunday and why, despite its life-saving promise, the federal government was slow to fund embryonic stem cell research under the Administration of George W. Bush. Separation of Church and State in the US may never be anything more than a political ideal due to the presence of religious interest groups. Such groups want the world around them to mirror their spiritual ideals and are willing to pay, in money and political effort, for such an outcome. In Indiana, for example, political power is sometimes used to shroud public spaces in religious propaganda. In its majority, the Indiana legislature has been known to view the *Ten Commandments* as "fundamental principles that are the cornerstone of a fair and just society" and to support their display at public buildings. Finding it difficult to sell this viewpoint to the federal government, Hoosier policymakers have shifted their political agenda to a smaller but more prevalent public venue. This latest example of spiritual propagandizing comes in the form of the state license plate.

Some specialty plates in Indiana bear the words, "In God We Trust." This statement seems innocuous enough—unless, of course, one doesn't hold the mainstream religious views of the state. According to recent data from *Pew Research Center*, 26% of Hoosiers are not affiliated with any organized religion [1]. Not without a sense of humor, *Pew* calls these individuals *religious nones*. When viewing such specialty license plates, some dissident *Hoosiers*—religious nones that they are—might wonder: *In which God is my state government trusting exactly, and why is this God so interested in the blanket trust of Midwestern humans? So*

S. Sanders, *The Economic Reason*, https://doi.org/10.1007/978-3-030-56043-0_7

interested, in fact, that the word "trust" is capitalized in the slogan. Certainly, religious propagandizing in the *Hoosier* state rates low on the all-time religious persecution scale. If modern-day California is a 0 and early 1940s Germany was a 100 on this scale, then Indiana, as it currently stands, is orders of magnitude closer to 0 than to 100. Still, the behavior of Indiana legislators is enough to make some religious nones in the Hoosier state wonder: *Is my state government respecting my religious beliefs or, rather, propagandizing a contrasting set of beliefs for the purpose of thought coercion?*

Despite their status as specialty plates, these small billboards cost no more than the standard Indiana license plate. To purchase something from the state's line of "Godless" specialty plates, however, one must pay an extra charge. Such an arrangement represents a de facto government subsidy to those who wish to express a particular set of spiritual views. Considering the arrangement from a slightly different perspective, it is a tax on members of specialty plate groups from which an ostensibly "Godful" group is protected.

It is not my contention that religious groups are wrong to have a public agenda. Everyone is free to compete in the political arena. Religious lobbying is largely successful because, over many moral issues, religious groups are more organized and place a higher per capita value upon political influence than do other competing parties. However, one might ask why it is not enough for a religious group to maintain its own moral standard and leave everyone else alone? Why do many such groups force the government's hand on issues such as abortion, for example? This question can be answered by exploring the nature of a belief. An academic study by Thierry Mayer, Anne-Celia Disdier, and Keith Head [2] finds that a person is more likely to hold an unconfirmed belief as the belief's popularity grows around her. Similarly, a person will have *more conviction* in a held belief as it becomes more popular. Religious belief is certainly no exception to this finding, as noted in a related study by Kenneth Scheve and David Stasavage [3, 4].

In economic terms, *Faith* can be viewed as a good that is subject to network externalities. Network externalities exist for any good that becomes more valuable as additional people adopt it. For instance, the fax machine and Internet are subject to network externalities. The first Internet connection wasn't useful in and of itself, but the second and third connections increased the value of the first substantially and profoundly. Later on in the history of the Internet, your connection might have become only slightly more valuable when your Aunt Estelle finally got a connection. *Faith* appears to have network externality characteristics in that a belief gains empirical credibility as it becomes more common. During the course of history, then, several religious groups have recruited fervently and attempted to push non-conformists into submission by making nonconformity illegal or at least psychically costly. As the Earthly world more closely approximates a religious group's spiritual ideals, *Faith* becomes more valuable to members of the group.

Legislation, the contract by which societies agree to live, is continuously negotiated. As with more traditional forms of government, a democratic government is perfectly capable of oppressing members of its citizenry (see, for example, much of US history). After all, elected representatives can be influenced just as dictators.

As contracts, documents such as the *US Constitution* are worth no more than the paper on which they are written unless properly understood, interpreted, and applied to present societal issues.

The Role and Scope of Economic Regulation

What, exactly, is the purpose of economic regulation? For many decades, the prevailing belief was that governments impose regulatory policies to increase the welfare that consumers derive from markets. In 1971, the late *University of Chicago* economist George Stigler wrote a seminal paper on the economic theory of regulation [5]. Largely for this work, Stigler won the 1982 Nobel Prize in Economics. Stigler's theory of regulation holds that regulation does not necessarily seek to improve consumer welfare or even the total welfare of society as a whole. Instead, economic regulation is "acquired by the industry (through political lobbying) and is designed and operated primarily for its benefit." There now exists a great deal of evidence in favor of Stigler's theory. For example, regulatory economist Robert Crandall has pointed out that the regulatory practice of holding interest rates on consumer checking accounts to zero could not possibly act toward the benefit of consumers.

In his seminal 1971 study, Stigler points out that oil import quotas, which at that time drove up the domestic price of oil substantially, could not have been implemented with anyone but the domestic oil producer in mind. Along these same lines, large cities often sell a limited number of taxicab medallions. This practice represents the creation (and capture) of monopoly rents by municipal governments. Of course, consumers lose in that they must pay a higher price and endure a restricted output of taxicab service. It is no surprise that *Uber* and *Lyft*, both of which have sidestepped taxi regulations in the United States, quickly relegated cabs to the status of a second class service in recent years. Anticipating this threat, many cities created bureaucratic market entry hurdles for *Uber* and *Lyft* as the companies sought to expand throughout the United States. The imposition of such hurdles further demonstrates the largely anti-consumer nature of economic regulation in the United States.

"New source biased" regulations are those that apply strictly to new firms of a particular industry. For example, the *United States Clean Air Act* of 1970 applied strict pollution standards to industrial firms that were new at the time, while exempting many older firms of such industries [6]. New source biased regulation is extremely inefficient in that it stifles industry incentives to build new plants. In this sense, the 1970 iteration of the *Clean Air Act* may have had the net effect of dirtying our air. Despite its overall inefficiency, Crandall finds [7] that such regulation was in the interest of old manufacturing firms of the Eastern United States at the time of the legislation. That is, it served to slow the migration of manufacturing jobs from unionized, Eastern states to nonunionized, Southern states. Thus, new source biased regulation arises when established firms of an industry have a strong interest in the creation of artificial barriers to entry and pursue that interest by wielding political influence.

Such bizarre regulatory actions occur on the local level as well. When my sister was a college student, a municipal law passed in the town of her university forbade smoking in newly opened restaurants and bars, while leaving existing establishments alone. Of course, the law had much more to do with protecting established businesses than with providing a smoke-free environment to consumers. Indeed, only a couple of restaurants opened during the years that this artificial barrier to entry was in place.

References

1. Pew Research Center. (2020). Religious landscape study: Adults in Indiana. Retrieved from https://www.pewforum.org/religious-landscape-study/state/indiana/
2. Disdier, A. C., Head, K., & Mayer, T. (2010). Exposure to foreign media and changes in cultural traits: Evidence from naming patterns in France. *Journal of International Economics, 80*(2), 226–238.
3. Scheve, K., & Stasavage, D. (2005). Religion and reform: The political economy of social insurance in the United States, 1910-1939. *University of Michigan, mimeo.*
4. Scheve, K., & Stasavage, D. (2006). Religion and preferences for social insurance. *Quarterly Journal of Political Science, 1*(3), 255–286.
5. Stigler, G. J. (1971). The theory of economic regulation. *The Bell Journal of Economics and Management Science,* 3–21.
6. Levinson, A. (1999). Grandfather regulations, new source bias, and state air toxics regulations. *Ecological Economics, 28*(2), 299–311.
7. Crandall, R. W. (1983). Clean air and regional protectionism. *The Brookings Review, 2*(1), 17–20.

Restricting the Marketplace

8

Many markets face some form of government intervention.
This is sometimes due to the presence of market failure and
sometimes due to the equity concerns of policymakers. In this
chapter, we consider the many effects of market intervention
by governments.

Price Controls

At times, governments pass legislation to regulate the price of a good or service. There are two main forms of price regulation: a *price floor* and a *price ceiling*. Far from being dry economic concepts, price regulations can profoundly influence "who gets what and why," a compound question coined by Nobel Prize winning economist Alvin Roth to describe the outcomes of both regulated and free marketplaces [1]. Often and increasingly in our world, markets represent the allocative mechanism that sorts out these issues. In turn, price movement within markets is central to this sorting.

A price ceiling is a policy that sets a maximum legal price at which a given good or service can be sold. A ceiling price of $1000 states that the producer *shalt not sell a unit for more than $1000*. In such a case, $1000 is the ceiling of all unit sale prices available to the firm when transacting with consumers. A price floor is a policy that sets a minimum legal price at which a given good or service can be sold. A floor price at $11.75 states that the consumer of a good or service *shalt not buy a unit for less than $11.75*. In such a case, $11.75 is the floor of all unit purchase prices available to the buyer when transacting with firms.

Price regulations are typically designed to help one "deserving" side of a market—consumer or producer—over the other. However, the road to poorly functioning market was paved with good intentions. Price regulations adversely affect the functioning of competitive markets. These effects include restricting the volume of market exchange and compromising the *allocative efficiency* of the market for units

S. Sanders, *The Economic Reason*, https://doi.org/10.1007/978-3-030-56043-0_8

that are exchanged. In the long run, price regulations can also lead to second-order effects such as quality degradation of the good being exchanged. In some cases, these unintended consequences are sufficiently strong that price regulations make *both* sides of the market worse off in terms of how much total value these parties derive from market exchange.

A market unencumbered by regulation is known as a *free market*. While free markets are often a good thing for society, this is by no means always the case. Specifically, markets that feature some form of market failure often need intervention in order to function properly. In a free market, price and quantity are not constrained from attaining their natural equilibrium allocations, wherever those might be at any point in time. Price is the common rationing mechanism of a free market, and units of a good are allocated to those consumers willing to pay at least the equilibrium price of the good. The equilibrium price has the characteristic that it creates a binary distinction between buyers of a good: *If you are willing to pay at least the equilibrium price, you get a unit. If you are not, you don't.* This characteristic is known as the market-clearing property of equilibrium price. At the equilibrium price and at no other price in a standard market, the quantity that consumers wish to buy equals the quantity that producers wish to sell.

The market-clearing property of the equilibrium price is no great surprise: equilibrium price occurs at the unique intersection point of supply and demand. At this price point alone, we can see that price is such that both sides of the market agree upon the quantity of units to be exchanged. Hence the equilibrium. In general, an equilibrium is simply a self-reinforcing or self-perpetuating allocation. Once you get to that outcome, it tends to persist. In any standard market, there exists one price at which consumers and producers agree upon the number of units that should be exchanged. It is the single-cross property of a standard supply and demand model that ensures a unique equilibrium point.

A Ceiling on Prices

Let us consider the effects of a price ceiling in greater depth. This policy is typically enacted in an attempt to help consumers purchase units of the targeted good at an attainable price. That is, it is intended as a pro-consumer policy. For example, the price of a good in a free-market setting may be prohibitively high for many consumers due to such factors as short supply, strong demand, or market power among firms provisioning the good. Within such a free market, it is often the case that relatively few prospective consumers are willing to pay at least the equilibrium price to buy the good. That is, only a small fraction of those valuing the good might actually be allocated a unit under free-market equilibrium. For example, real estate in Manhattan is scarce because Manhattan is a small but happening island. The *small* part dictates a short supply of real estate on the island, while the *happening* part dictates a strong demand for the same.

Given these conditions, a small proportion of those who wish to live in Manhattan actually do so. For every *bona fide* Manhattanite, there are dozens of wannabes hoofing and ferrying their way to Manhattan hotspots each day from places like Queens and Jersey. Such market outcomes can evoke feelings of unfairness among those consumers not served under the market equilibrium allocation. These feelings can resonate with municipal legislators, sometimes leading to the enactment of rent controls—price ceilings on apartment rental prices. In fact, rent controls have a long history of use in Manhattan. A recent period of this history is memorialized in the popular 1990s television show *Seinfeld*. The show profiles several of the effects of a rent control policy, as observed from the perspective of a Manhattan apartment renter [2].

This element of the show is discussed in an economics pedagogy paper that I co-authored with fellow economists R. Andrew Luccasen and Abhinav Alakshendra [3]. In the paper, my co-authors and I list the types of welfare costs imposed by rent control, as profiled in episodes of the show. Specifically, the show highlights the difficulty of apartment procurement under rent control (shortage, tastes for discrimination on the part of landlords, bribery, search costs) and failure on the part of the landlord to keep up the quality of apartment units over time (quality degradation). *Seinfeld* also illustrates the informal process through which rent-controlled apartments are advertised (less advertising under rent control shortage). The paper states,

> This paper analyzes four episodes of *Seinfeld* to help students identify and differentiate the very real costs and conditions of rent control. The paper also guides students to appreciate the difficulty in crafting a policy that is free of unintended consequences...If *Seinfeld* is about anything, it is about suffering (i.e., cost bearing) at least in the urban, middle-class, American sense of the word (p. 1).

The Andrea Doria episode demonstrates that shortages of a good arise when the good's price is not allowed to rise to the free-market equilibrium price. In this case, the price mechanism cannot move freely to ration units of the good and to balance the valuations of buyers and sellers. As a result, alternative or non-price rationing mechanisms must be employed. In this episode, George vies for an apartment against other prospective renters. In other words, quantity demanded for the apartment exceeds quantity supplied at the existing rental price, which serves as an indication of rent control. As a non-price rationing mechanism, the tenant board considers the extent to which each potential renter has experienced personal challenges in life. They do this to determine which candidate is "most deserving" of the unit. An excerpt on page 6 of the paper states,

> George is notified by the tenant association that the apartment will be given to a survivor of the *SS Andrea Doria* shipwreck "in light of his terrible suffering." George and Jerry have the following exchange regarding this non-price mechanism:
>
> *George*: So, he's keeping the apartment. He doesn't deserve it, though. Even if he did suffer, that was, like, 40 years ago! What has he done for me lately? I've been suffering for the past 30 years up to, and including yesterday!
> *Jerry*: You know, if this tenant board is so impressed with suffering, maybe you should tell them the "Astonishing Tales of Costanza."

George: I should!
Jerry: I mean, your body of work in this field is unparalleled.

In the end, the tenant board decides to allocate the apartment unit on the basis of a bribe payment, revealing another non-price rationing mechanism under rent control. While landlords cannot directly charge a monthly rental price that exceeds the allowable rent-controlled price, they have been known to accept an upfront payment that compensates them for the reduction in monthly rental price. Bribe payments for rent-controlled apartments act to nullify the intended effect of a rent control policy. Let us assume, for example, that an apartment unit has a free market equilibrium rental rate of $2500 per month. Let us further assume that the apartment is rent-controlled at $2000 per month. The apparent effect of the policy is to reduce the landlord's 12-month receipts by $6000. We know, however, that there is one consumer willing to pay $30,000 for the right to live in the apartment for a year. *How do we know this?* We know this because $2500 per month was defined as the equilibrium monthly rental rate. An equilibrium rate leads to market-clearing conditions, whereby quantity supplied equals quantity demanded at that price. As such, there is exactly one buyer of this apartment at the equilibrium rate of $2500 per month ($30,000 per year) *by definition*. If the landlord is open to bribe payments, then this single renter is expected to pay the highest bribe of any potential renter of the apartment. Namely, she will pay up to $6000 on top of her monthly rental rate. In doing so, the renter has nullified the rent control policy.

Bribe payments recur in the *Seinfeld* episode titled *The Apartment*. In that episode, an older tenant in Jerry's apartment building passes away, and Elaine is immediately interested in renting the vacant apartment. However, the landlord expects a bribe payment that Elaine cannot afford. Elaine's interest in the apartment is not quelled by the circumstance that creates its vacancy. Such vulture-like behavior is a common response within a setting of rent control. Under shortage, renters incur substantial search costs to find an apartment. In such a climate, scruples and shyness are dear luxuries that renters cannot afford. Residents of Cedar Rapids, Iowa, or Peoria, Illinois, are unlikely to vulture an apartment from a dead neighbor or to offer an upfront bribe payment when attempting to rent a place. Does this say something about the integrity of the residents in these Midwestern cities vis-à-vis that of New Yorkers? The economist would say that there is no accounting for preferences concerning scruples. On the other hand, there is accounting for differences in incentives.

Vast expanses of farmland surround Cedar Rapids and Peoria on all sides. A new house can be built on this land for not much more than the price of farmland. Existing rental units are readily available in these areas, as well. Moreover, isolated Midwestern cities not named *Chicago* typically hold very little cultural or social significance for those not from the region. In a 2010 interview, former *NBA* player Joakim Noah offered a quote about the city of Cleveland that may apply equivalently to Cedar Rapids or Peoria. When asked if he regrets previously deriding the city, Noah stated, "Not at all. You like it? You think Cleveland's cool. I mean...I never heard anyone say 'I'm going to Cleveland on vacation.'"

Surrounding Manhattan, on the other hand, is water on all sides. Manhattan is an island that possesses cultural, social, and economic significance on a global scale. If you wish to join this highly demanded ecosystem by building a new home, it is not possible to simply buy a little farmland and nail together a nice little (one-story) ranch home. For the most part, the only place to build in Manhattan at this point is up...way up, and it is quite expensive to do so. You tend to need a lot of steel in order to build houses upon other houses many times over. Manhattan's characteristics, short supply of living space and strong demand for said space, create a perfect storm for housing scarcity. Aggressive apartment search behavior may be closer to a survival instinct for devoted Manhattanites than for those living in cities that are more like Cedar Rapids. In the big, rent-controlled city, she who flinches may end up in Staten Island.

People respond to incentives. As such, it is very likely that scarce housing is a chief driver of apartment search behavior in Manhattan. Scarce housing, however, is not synonymous with a housing shortage. A housing shortage represents an even worse state of affairs. Shortages occur when prices cannot move freely up to the equilibrium level. As a result, even fewer people are served under a rent-controlled market because landlords have less incentive to provision rental apartments. Under the rent-controlled price, some landlords may decide to convert apartment units into office space or, in the long run, to sell old apartment buildings wholesale to developers of commercial real estate. This is the chief unintended consequence of price ceilings: *they constitute a tradeoff rather than a gain for the group they seek to protect*. Namely, consumers who are lucky or resourceful enough to obtain a unit of the good under shortage receive the unit at a lower price. However, fewer people are served at the rent-controlled price because quantity supplied of the good has declined.

At the same time that quantity supplied declines, quantity demanded increases at the lower price such that a shortage is created from both sides. At the equilibrium price, quantity supplied is equal to quantity demanded, allowing a competitive market to attain allocative efficiency and maximum economic surplus. When the price mechanism is cut off from attaining its equilibrium level due to a ceiling on prices, the efficient market allocation is no longer possible. For every q_d units that are "demanded" at a binding ceiling price, only q_s of these demands will be met by willing suppliers, where $q_s < q_d$. Given this predicament, a binding rent control creates a lottery element to apartment searches. As the number of apartments available is fewer than the number of people searching for an apartment at any sub-equilibrium price, apartment hunters incur substantial search costs in Manhattan and other rent-controlled cities. Even then, not all individuals willing to pay the rent-controlled price will find an apartment under such conditions.

One can visualize this consumer predicament. Imagine a long line of consumers waiting for a good that, for whatever reason, is not priced at equilibrium level. This good may be price controlled, or it may be priced differently due to a larger promotion that a store is running. On Black Friday, for example, retail stores such as *Best Buy* tend to price down certain popular electronic items for the purpose of getting people in the store. There is no advertising, after all, like feet in the door. We can think of such a promotional sale as a sort of self-imposed price ceiling (i.e., one levied upon the store by the store itself). Let us say that a *Best Buy* retail location prices an *Xbox* at a sale price of $150 on Black Friday. The catch is that the store stocks only 100 units of the item and limits purchases to one unit per customer. Let us further imagine that there are 300 people waiting in line for the purpose of buying a unit. In this spot market, quantity demanded is 300 units, and quantity supplied at the sub-equilibrium price is 100 units. Therefore, only one-third of those willing to pay the posted price will obtain a unit. At equilibrium price, on the other hand, all consumers willing to pay the posted price receive a unit. Once again, this latter outcome is referred to as the "clearing property" of the market equilibrium price.

Under normal circumstances, market clearing is visible in retail settings. Namely, we see it in the form of fairly constant inventories within the retail stores that we frequent. When you go to the grocery to buy a box of your favorite fruit cereal, it is very unlikely that the store has stocked out of the cereal. It is also unlikely that the store managers purchased so many boxes of the cereal that they have run out of shelving and back storeroom space. Rather, the number of boxes in stock is likely fairly steady over time. Whereas some of this outcome is due to supply chain management technologies, much of it is due to the role of equilibrium pricing in balancing the needs of consumers and producers in the marketplace. If the cereal were placed on sale, it may stock out. If the price were doubled, boxes of the cereal might collect noticeably on the grocery store shelves. At the equilibrium price, however, the market clears and inventories are largely balanced.

At equilibrium price and quantity, and at no other allocation of price and quantity, there is exactly one unit supplied for each unit demanded. If the market price is held below equilibrium, then there will be more than one unit demanded for every unit supplied. Thus, sub-equilibrium pricing creates a shortage. In the *Best Buy* example, we know exactly who will obtain units of the good. Namely, it will go to the first 100 *Xbox* seekers who are in the line. Market outcomes are not always so clear-cut, however. When price cannot reach equilibrium to ration units of a good, queuing is just one form of non-price rationing that determines who gets what in the market-place. In the case of rent-controlled housing, luck and costly search efforts serve as pivotal inputs in allocating apartment units within the rent-controlled district.

We might view the expenditure of costly search or queueing efforts by consumers as something that leads to a sort of meritorious allocative process: *Those who wait in line the longest or who search for an apartment most diligently have a better chance of being allocated a unit of the good.* After all, hard work is commonly viewed as a virtue. In this particular case, however, an economist may have some difficulty accepting that effort-based allocation is any more desirable than lottery-based allocation when rationing a price-controlled good. Queuing and search efforts to procure a price-controlled good represent a sort of work that is productive for the individual but not so productive for society as a whole. Such behavior falls in a class of behaviors known as *rent seeking*, where *rent,* in this case, is defined as economic gain.

In general, rent-seeking behavior creates little or no value from the perspective of society [4, 5]. Rather than creating a good or service, the rent seeker works to effect a transfer of economic surplus from others to herself. In this sense, rent seeking is close to a zero-sum activity, whereby the winner's gain is the loser's loss. There may be some allocative efficiency gains from certain forms of rent-seeking behavior. After all, those willing to incur the highest search or queuing costs for a unit of a good are often those who hold a relatively high valuation for the good. One might counter, however, that those willing to wait in a queue may simply have a lot of time on their hands. Overall, the allocative implications of search contests are therefore unclear. Regardless of these smaller considerations, rent-seeking behavior is, for the most part, economically costly. Rather than putting forth efforts to create value on a social scale, rent-seeking individuals put forth efforts that have the primary effect of influencing the transfer of a resource that already exists.

Rent-seeking behavior has analogies to behavior in sports competitions, especially at the youth level. In youth basketball games, there are often two ways that players request a pass. A player can put forth effort to get open and create a better opportunity for her team to score. This would qualify as productive effort, as the player is attempting to move the ball to a part of the floor that is less heavily defended and from which it is otherwise easier to score. Another common approach at the early youth level is for an offensive player to crowd her ball-handling teammate and strongly request a hand-off pass. The hand-off pass is not really productive, as it does not advance the team's overall scoring agenda. Rather, it simply replaces one well-defended, stationary ball handler who is usually far from the basket with another. Requesting a hand-off pass in basketball is akin to rent-seeking behavior. By crowding the ball handler, a teammate is trying to effect a transfer that has value to the individual—young players usually believe that touching the ball is an accomplishment in itself—but no real value to the overall team goal.

An efficient team attempts to promote incentives that are in the best interest of society. Teams usually accomplish this goal through coordinated efforts (e.g., set plays) that are managed by a coach, a point guard, or a team captain. Similarly, free markets can often align individual and social interests through a freely adjusting price mechanism that ensures allocative efficiency.[1] Like a good basketball coach, price adjustment acts to balance individual interests toward a social goal. In the case of market exchange, the goal is that the market creates the most possible net economic value.

---

[1] A free or unregulated market does not always represent a socially optimal form of allocation for a good. Rather, this is the case provided that a good is competitively provisioned and competitively bought. If there are too few suppliers, we have oligopoly market power on the supply side. If there are too few buyers, we have oligopsony market power on the demand side. If market power exists on either side of a market, free market allocation is no longer efficient. In such a case, a freely adjusting price mechanism is like a poor basketball coach in that it can actually coordinate away from the social optimum.

Back to the Long Run

As was suggested in a previous passage of this chapter, landlords have additional options by which to ration rent-controlled apartments in the long run. Namely, they can simply cease to maintain them. Failure to keep up a rent-controlled apartment will lead to quality degradation and, in turn, demand decrease in the long run. This inward demand shift reduces the gap between the rent-controlled price of an apartment unit and the going market price such that the level of apartment shortage declines *because of* quality degradation. While Manhattan is a coveted place to live, fewer people will want to live in an aging dump there than in a standard apartment. Instead of scouring Manhattan for a home, more individuals will move to a relatively spacious and well-maintained place in Jersey or other surrounding areas as the average quality of Manhattan apartments declines. Once quality degradation took root in rent-controlled Manhattan apartments, it follows that daily ferry commutes to the island no longer appeared to be so undesirable by comparison. Quality degradation is another recurring rent-controlled apartment theme in *Seinfeld*. As the show ran for a full nine seasons, with Jerry in the same aging apartment throughout, the long run eventually set in for him and his apartment. In the long run, not all shows are dead.

In some cases, quality can degrade abruptly in response to a decision by the landlord. *Seinfeld* chronicles an example of such a form of quality degradation in the episode *The Shower Head*. In this episode, Jerry's landlord installs low-flow showerheads without consulting any of the renters. After these new showerheads are installed, the renters feel that their landlord has lowered the quality of the apartments. Let us examine how quality degradation can help the landlord of a rent-controlled apartment unit. Consider an apartment unit that has a free market equilibrium price of $2500. Let us further assume that the landlord must spend an average of $500 per month on the upkeep of the unit to rent out the apartment in a free, competitive market environment.[2] In this scenario, the landlord averages a net receipt of $2000 per month in exchange for renting the unit. Let us imagine that a rent control policy is suddenly passed in the very city district that this apartment is located. Due to the policy, the landlord can now charge no more than $2000 for the apartment. The apparent effect of this policy is that the landlord now nets ($2000 − $500) = $1500, and this may, in fact, be her short-run rate of gain. For example, there is a chance that the landlord is in the middle of financing a renovation or an appliance swap when the policy is passed. Such improvements will cease in the long run. The landlord will quickly realize that the market no longer clears. Instead of one renter for her apartment at the equilibrium price, there are now perhaps two or three potential renters each of whom has no other immediate options in the neighborhood. What is the landlord to do?

---

[2]We represent upkeep as an average cost because many of these expenses can occur in the long run (e.g., renovations, utility and appliance replacements,. . .).

If she is like most profit-maximizing landlords under regulatory duress, she is to continue only with upkeep that is necessary to a) meet city occupancy ordinances and b) keep at least one renter interested in the apartment. After rent control is imposed, let us say that the landlord spends just $50 per month on the upkeep of the unit. Over time, that's going to leave a mark. In following this cut-rate approach, however, the landlord is able to recoup some of her monthly losses from rent control. She is able to transfer some of these costs back on to the renter, as the renter no longer values living in the apartment by as much. Instead of netting $1500 per month, the landlord nets $2000 − $50 = $1950 per month for the unit. In the long run, her net receipts have declined by only $50 per month as a result of the rent control policy.

It is not that consumers fail to notice the quality degradation of the apartment. As noted earlier, demand declines as consumers move out to neighboring boroughs and cities. The resulting decrease in demand may eventually cause only two instead of three potential renters to queue onto the apartment's waitlist. As the apartment becomes further dilapidated, demand may further slide until there is no longer a waitlist at all. At this point, quality degradation has decreased demand by so much that the rent control policy no longer binds, and the market attains a sort of low-level equilibrium in the long run. At this low market level, the equilibrium price has fallen until it is just equal to the rent control price. However, it should be recognized that the nature of the market has fundamentally changed at this point due to the imposition of the rent-control policy. Whereas the market used to allocate a presumably prime apartment in a prime location, it now allocates a Grade A dump in a prime location. Rent-controlled apartments may represent the chief impetus for New Yorkers to spend so many hours at the office or at Central Park. Indeed, rent control can be costly to consumers in the short run but even costlier in the long run.

Price Ceilings Away from Home

We can consider in what other settings price ceilings are present. Laws against so-called *price gouging* impose a ceiling price on necessity goods in the wake of a natural disaster [6]. During hurricanes, for example, the demand for bottled water rises precipitously, as freshwater supplies become compromised. At the same time, it becomes more costly to supply bottled water to hurricane-stricken areas during such events. It isn't always so simple as driving a large number of delivery trucks to stores in the region. For example, the roads leading into the affected area may be washed out. Thus, the supply curve for bottled water and other goods decreases following many forms of natural disaster. When demand rises and supply decreases, prices can rise tremendously if left to market forces. In 2017, Hurricane Harvey inundated coastal areas of Texas. In its aftermath, cases of bottled water reportedly rose to as high as $99 and gasoline prices to as high as $20 per gallon in affected areas of Texas [7].

Though consumers normally accept market price as a mechanism to balance market interests, social norms can override market norms in times of emergency.

When the price of a necessity rises precipitously at an already challenging time, it can feel like being punched when one is already down. In such cases, irate consumers and government representatives sometimes seek retribution. For example, many Texas consumers filed price gouging complaints in the aftermath of Hurricane Harvey. While the term *price gouging* is not well-defined, many consumers and legislators feel as though they know an act of price gouging when they see it. Laws against "price gouging" seek to help consumers in times of need, but do these laws actually meet this goal? On the surface, it seems as though such laws allow individuals to afford necessity goods in times of emergency. High price is not the only market element regulating access, however. As discussed previously, sellers restrict quantity supplied in response to a binding price ceiling.

While it may seem that sellers would not restrict quantity supplied of bottled water in response to, say, a $50 per case price ceiling, we must remember that the economics of natural disasters can be quite bizarre. Some producers may themselves pay exorbitant distribution fees to restock inventories following a natural disaster. Others may have paid substantial storage costs to stock goods for a long period of time in anticipation of a disaster. Some small-scale suppliers may be deciding how much they value having the case of water for themselves, their family, or their friends. As such, the opportunity cost of selling a case of water may be very high in the wake of a natural disaster. Thus, price gouging laws may in fact reduce access to bottled water and other essential supplies. Economic analysis often yields that the obvious solution is not a solution at all.

Following natural disasters, there are other market-related policy options that unequivocally improve the expected outcomes of affected individuals. In cases of severe, anticipated natural disasters such as a Category 4 or 5 hurricane, mandatory evacuations may improve outcomes.[3] While evacuations do impose a cost on individuals, they also mitigate the potential worst-case costs associated with natural disasters. They do this partly by holding in check demand for essential services and supplies in the disaster region. Without a large population of victims in the affected region, the price push for essential goods is expected to be less aggressive and perhaps substantially so.

Direct subsidies and government provisions of essential items are other types of government policies that tend to be effective toward disaster relief. In some cases of disaster, governments step in to provide and distribute goods. Public subsidies and provisions serve to counteract the dramatic inward supply shift associated with many natural disasters. As with mandatory evacuations, then, these policies have the effect of limiting the rise of essential goods prices following a natural disaster. In this sense, government subsidies and provisions have the same qualitative price effect as anti-gouging price ceilings. Unlike price ceilings, however, government subsidies and provisions improve access to units of essential goods rather than decreasing said access. Government subsidies and provisions also hold certain advantages over

---

[3]Mandatory evacuations, when issued, usually affect all citizens who are non-essential to the disaster management and relief effort.

mandatory evacuations. While evacuations require the benefit of foresight, government provisions and subsidies can be reactive in nature. In this respect, subsidies and provisions by government are applicable to a broader range of natural disaster events.

A Floor on Prices

The question as to whether laws against price gouging are more effective than direct government interventions is of vital importance in the wake of disaster. There is an analogous discussion that occurs in low-skill labor markets, in which governments can either impose a minimum wage law or subsidize employee pay rates to effect wage increases in these markets. A minimum wage represents a price floor. Once instituted, it is illegal to pay workers below the minimum wage. A wage subsidy is a stimulus that acts to increase employer demand for workers. As with minimum wage laws, wage subsidies are typically targeted toward low-skill, low-wage workers. They work as follows: For every hour of low-skill labor that a firm hires, the government pays $X to the employer in support of paying the worker's wage. Effectively, wage subsidies increase firm demand for low-skill workers. Like a binding minimum wage law, a wage subsidy increases the wage rate paid to these workers. Unlike a binding minimum wage law, a wage subsidy also acts to increase the number of hours these workers are employed. Wage subsidies support higher wages *and* higher employment levels.

As with binding price ceilings, binding price floors distort prices and create a gap between quantity supplied and quantity demanded in a market. Whereas price ceilings cause shortages—outcomes in which quantity demanded exceeds quantity supplied—price floors create surpluses. Surpluses can be thought of as the allocative opposite of shortages. That is, they are cases in which quantity supplied exceeds quantity demanded. To understand why a surplus arises under binding price floor, let us imagine a fish market. This market is operating smoothly until, one day, the city council imposes a binding price floor on fish sold within the market. Specifically, they say that no fish can be sold for less than $15, whereas the equilibrium market price is $10. The policy is designed to help fishermen, who complain that it is difficult to make a living on $10 of revenue per fish. Responding to these stronger incentives, fishermen catch and deliver to market more fish than ever before. However, several consumers balk at the new price of fish and substitute chicken and goat meat in its place. The market does not clear at the floor price of $15 per fish. In the case of a fish market, a surplus will be quite memorable. After all, human memory is strongly linked to the olfactory sense! [8]. We have seen previously that binding price ceilings hurt producers and may not help consumers in the end. Conversely, binding price floors hurt consumers and may not help producers in the end.

Controlling Monopoly Prices

We have now summarized the unintentional consequences of price controls within a competitive market setting (i.e., one with many buyers and sellers, each of whom is well-informed within the market). Interestingly, the world we have described is flipped upside-down in a monopoly or monopsony market. A monopoly market is one that features only one seller, whereas a monopsony market is one that features only one buyer. A common example of a monopolist is a public utility (e.g., an energy provider) or the television broadcaster of the *Olympic Games* in the United States. In many geographic areas, there is only one electricity provider. Similarly, primary broadcast rights for the *Olympic Games* are sold to a single network in the United States. Examples of a monopsonist include the *National Football League* (NFL) or the *US Air Force* (*USAF*). The *NFL* is the only "buyer" of elite, professional American football labor, whereas the *USAF* is the only buyer of *B2 Stealth Bombers*.

Monopolists and monopsonists possess market pricing power, and they use it to increase profits. For simplicity, let us consider the case of monopoly markets in greater depth, where much of the analysis generalizes to the case of monopsony markets. Under monopoly provision, sellers restrict output to a number of units below the socially optimal level. They do so to raise the unit price and thus maximize profit by selling to the top portion of the demand curve (i.e., those with the highest willingness to pay). When a binding price ceiling is imposed upon a monopolist, the incentive to maximize profit by raising price is lost to an extent because the profit-maximizing price level is no longer available. The monopolist's alternative is to charge as high a price as is allowed (i.e., the ceiling price). At a binding ceiling price, which is lower than the market equilibrium price by definition, the monopolist will actually increase output as compared to the free market case. Whereas a price ceiling in a competitive market creates shortage and *reduces* exchange, this same policy in a monopolistic market decreases price and *increases* exchange. As such, binding price ceilings unequivocally help consumers in the monopoly market case. For this reason, price ceilings are often employed by governmental bodies to regulate so-called *natural monopoly markets*.

Just what is meant by the term *natural monopoly markets*? In some markets, it is "natural" to have a single firm do all of the provisioning. Specifically, monopolies are natural when one firm can produce for the whole market at a lower long-run average cost than can any two, three, four, or more firms. In the case of fiberoptic cable networks, for example, a large capital investment is required to lay the network. Let us imagine that, in a small city, two companies lay identical fiberoptic network cables side-by-side and split the local cable market. Each company's large capital investment is offset by units of cable subscription sold to residents of the city. However, each company supplies to only half of the market and therefore may bear a high long-run average cost per unit delivered. If long-run average cost is higher than price, then the two firms will be unprofitable. In a natural monopoly setting, it is often profitable for one firm to supply the entire market but unprofitable for more than one to do so. The very nature of physical costs in such markets reinforces monopoly provision.

In natural monopolistic markets, monopoly provision is a mixed bag. Natural monopolies do bring efficiency gains to the market in the form of lower long-run average cost. Some of these gains are passed on to consumers in the form of lower prices. However, monopolies also restrict output so as to raise price above that which would be charged in a competitive market. Further, natural monopolies do not have a lot of incentive to be "dynamically efficient" (e.g., innovative) in the long run, as they have a monopoly position that is being defended by nature itself. As market positions go, nature itself is a pretty good defender to have. In the case of most markets for which monopoly provision is natural, the United States and several other countries allow the firm to supply the market as a monopoly but restrict market allocations via price cap regulation. Price cap regulation is simply a price ceiling imposed on the regulated market. Given the nature of monopoly allocations, price cap regulations are imposed to encourage allocations in the market that more closely resemble a competitive market process. As suggested previously, price cap regulators can help consumers in two ways. In monopoly markets, price caps reduce price while also *increasing* quantity supplied and overall exchange. In turn, these effects help consumers achieve higher gains in a marketplace that would otherwise give them short shrift.

Whereas price cap regulations create net welfare losses in competitive markets, they create net welfare gains in monopolistic, oligopolistic, or monopsonistic markets. They do so by cutting off the equilibrium price and thus pushing firms to produce more units at a lower price. As such, optimal price regulation of public utilities, cable companies, airlines, and other firms that possess market power is necessary toward desirable market outcomes.

References

1. Roth, A. E. (2015). *Who gets what—and why: The new economics of matchmaking and market design*. Houghton Mifflin Harcourt.
2. Seinfeld is rich in economics examples, as demonstrated on: Ghent, L., Grant, A, & Lesica, G. (2020). The economics of Seinfeld. Retrieved from https://www.yadayadayadaecon.com/
3. Sanders, S., Luccasen, R. A., & Alakshendra, A. (2020). Rent control according to Seinfeld. Mimeo. Syracuse University.
4. Tullock, G. (2001). Efficient rent seeking. In *Efficient rent-seeking* (pp. 3–16). Boston, MA: Springer.
5. Tollison, R. D. (2004). Rent seeking. In *The encyclopedia of public choice* (pp. 820–824). Boston, MA: Springer.
6. Lee, D. R. (2015). Making the case against "price gouging" laws: A challenge and an opportunity. *The Independent Review, 19*(4), 583–598.
7. Garfield, L. (2017). $20 for a gallon of gas, $99 for a case of water: Reports of Hurricane Harvey price-gouging are emerging. Retrieved from https://www.businessinsider.com/price-gouging-in-texas-gas-prices-hurricane-2017-9
8. Willander, J., & Larsson, M. (2006). Smell your way back to childhood: Autobiographical odor memory. *Psychonomic Bulletin & Review, 13*(2), 240–244.

Bringing It All Back Home

9

This chapter brings together the themes of the book under the umbrella of economic thought.

This book has visited many different walks of life. But then, economics has a tendency to do that. In fact, economics is best used broadly. Given the power of the economic toolkit, the field has become defined more by its methodology and less by its subject area. At its best, economics is a general framework for understanding human behavior without qualifications upon the type of behavior studied. In a classic paper titled, "Economic Imperialism," [1] leading labor economist Edward Lazear discusses the flexibility and power of the economics paradigm. He states that the economics methodology makes it a social science in subject but a "genuine science" in methodology. Specifically, he asserts that economics uses and even develops statistical techniques rigorously to examine testable aspects of human behavior: efficiency, maximizing behavior, and equilibrium being chief among them. Further, the empirical tests that economists consider represent tests of an axiomatic theory of human behavior. Lazear concludes, "These ingredients have allowed economics to invade intellectual territory that was previously deemed to be outside the discipline's realm" (p. 99).

Indeed, economic theory departs fundamentally from the mainstream theoretical constructs of other social sciences. Conventional economic theory argues that stakes-based outcomes yield more predictive power than those not involving stakes. In game theory, for example, it is said that "talk is cheap." This means that people might say things they do not really believe when there is no stakes-based outcome associated with their words. Economists tend to be very wary of "cheap talk." It is a reason that the field is almost singular among the social sciences in its general disdain for survey questionnaires. For better or worse, a credo of economists might go as, "Ask not what individuals do. Observe what individuals do." Indeed, economists have little interest in autobiographical descriptions of behavior. People possess innate conflicts of interest when reporting their own behavior such that

S. Sanders, *The Economic Reason*, https://doi.org/10.1007/978-3-030-56043-0_9

reporting bias is invited into the data. Whereas Mother Teresa may have been expected to fill out a survey form truthfully, most of us wish to be viewed as more admirable than our behavior merits.

If talk and survey responses are cheap, how do economists obtain data upon which to test economic theories? How do they, in fact, "Observe what individuals do?" In the twentieth century, almost all economic data was obtained empirically or "in the course of events." In the mid-to-late twentieth century and into the twenty-first century, economists began to motivate stakes-based decision environments through economic laboratory experiments, field experiments, and simulated decision environments. A 2009 *Science* paper by Falk and Heckman [2] finds that experiments have steadily increased in the field of economics since being introduced in the late 1940s. The authors state,

> The first lab experiments in economics were not conducted until the late 1940s. Fewer than ten experimental papers per year were published before 1965, which grew to about 30 per year by 1975. Starting from this low level, experimentation in economics greatly increased in the mid-1980s. In three well-known economics journals—American Economic Review, Econometrica, and Quarterly Journal of Economics—the fraction of laboratory experimental papers in relation to all published papers was between 0.84 and 1.58% in the 1980s, between 3.06 and 3.32% in the 1990s, and between 3.8 and 4.15% between 2000 and 2008.

According to this historical characterization, the experimental methodology in economics progressed from non-existent to viable and then to oft-used in little more than a half-century's time. Today, more than 1 in 25 economics studies use this approach as its chief methodology. Part of this transition owes to computational advances that allow for laboratory experimental data to be collected seamlessly as it is generated. However, another part of this transition speaks to underlying changes in the thinking of economists. Economists have always sought to foster a heightening of the scientific method within the social sciences. For example, econometric methods have been developed to isolate individual relationships toward an estimation of causal pathways. Similarly, economists have found that experimental data can help to clean out many of the confounding factors that are implicit in most forms of naturally occurring economic data. In experimental contexts, economists typically introduce stakes by paying subjects according to their performance in the experiment, where the experiment typically represents a stylization of a real-world decision setting. These approaches have allowed economists to more definitively study traditional economic topics, while also meaningfully imperializing important social science topics that were previously outside of their domain.

During my days of economics doctoral study, I followed the philosophical path set by Lazear and others to develop a broad understanding of what economics is and what it can be. This view of economics was derived from several different sources: Lazear's article, the philosophy of economic thought held by my doctoral adviser, Dr. Yang-Ming Chang, and an innate interest in many different social science topics. My earliest economics scholarly work was shaped by this perspective, leading some colleagues to suggest that I wasn't actually "doing economics" at all. Their slight was lost on me, as I never cared to be defined by what mainstream economists before me had done. Even as a young economist, I had developed a bad case of economic imperialism. My prognosis worsened until, eventually, I took an academic position outside of an economics department. Since 2016, I have served as Associate Professor and then as Professor of Sports Economics & Analytics within the Sport Analytics program at *Syracuse University*. Am I still an economist? Was I ever an economist? A true economic imperialist might say the answer to each of these questions is an affirmative. Then again, a true economic imperialist would tend not to care for these questions in the first place. Labels and the idea of disciplinary boundaries tend to constrain, whereas the application of good methodology to new areas tends to free.

References

1. Lazear, E. P. (2000). Economic imperialism. *The Quarterly Journal of Economics, 115*(1), 99–146.
2. Falk, A., & Heckman, J. J. (2009). Lab experiments are a major source of knowledge in the social sciences. *Science, 326*(5952), 535–538.

Appendix: *R* Code for Trend World Battle Casualty Rate Plot of Chap. 6

```
library(tidyverse)

battle <- read_csv("PRIO.csv")
battle_year <- battle %>% group_by(year) %>% mutate(sumdead =
sum(bdeadbes))
population <- read_csv("population.csv")

summary(battle$year)
summary(battle$sumdead)
plot(density(battle$sumdead))

battleyear_pop <- inner_join(population,battle_year,by="year")
battleyear_pop <- battleyear_pop %>% mutate(casrate = bdeadbes/
population)

q <- ggplot(battleyear_pop,aes(x=year,y=casrate)) + ylab("World
Battle Death Rate by Year")
q <- q+ stat_smooth(method="loess",formula=y~x,size=1)
q

ggsave("battledeath rate.png")
```

CPSIA information can be obtained
at www.ICGtesting.com
Printed in the USA
LVHW011413300821
696438LV00002B/56